FAULT

RADIUS BOOKS

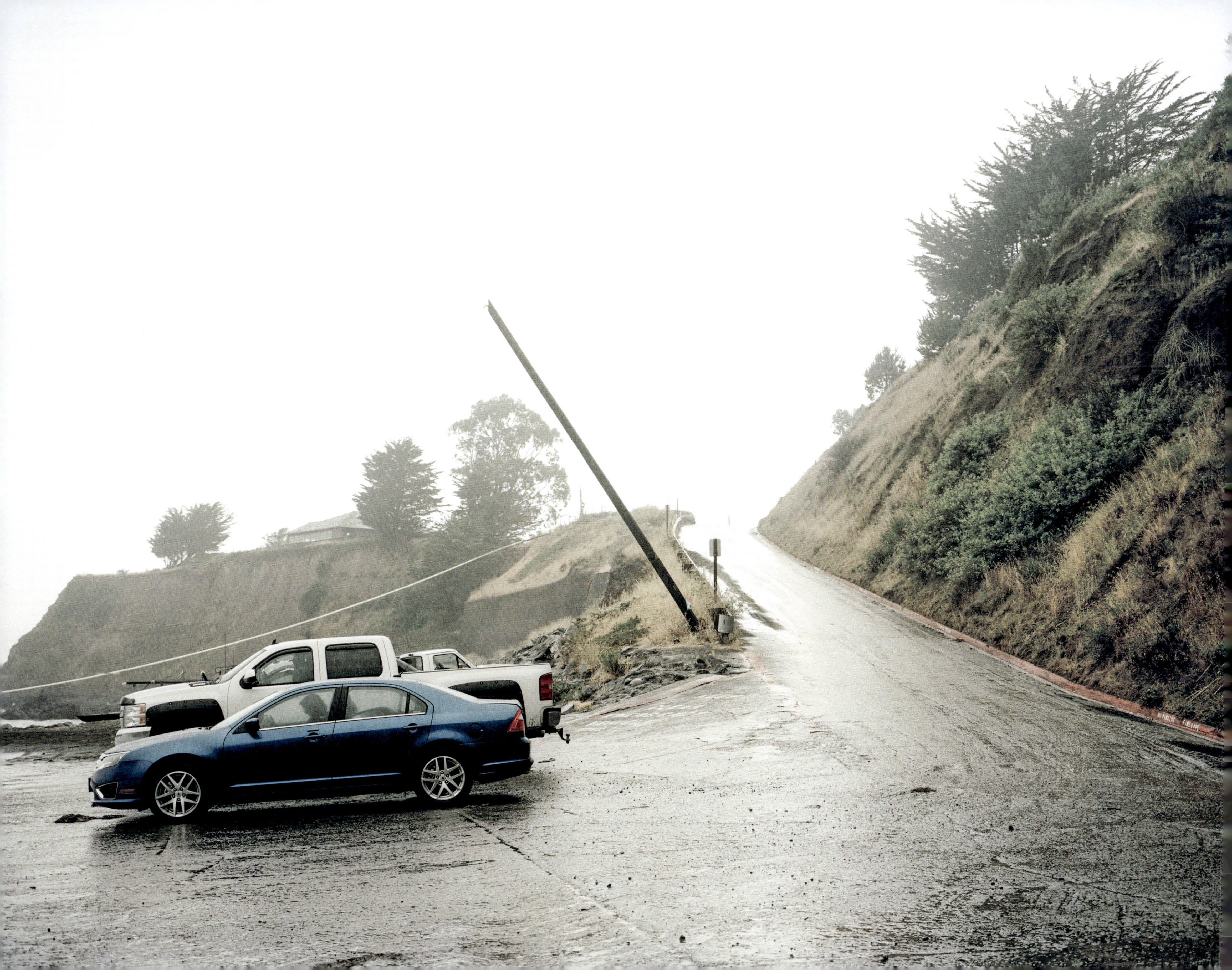

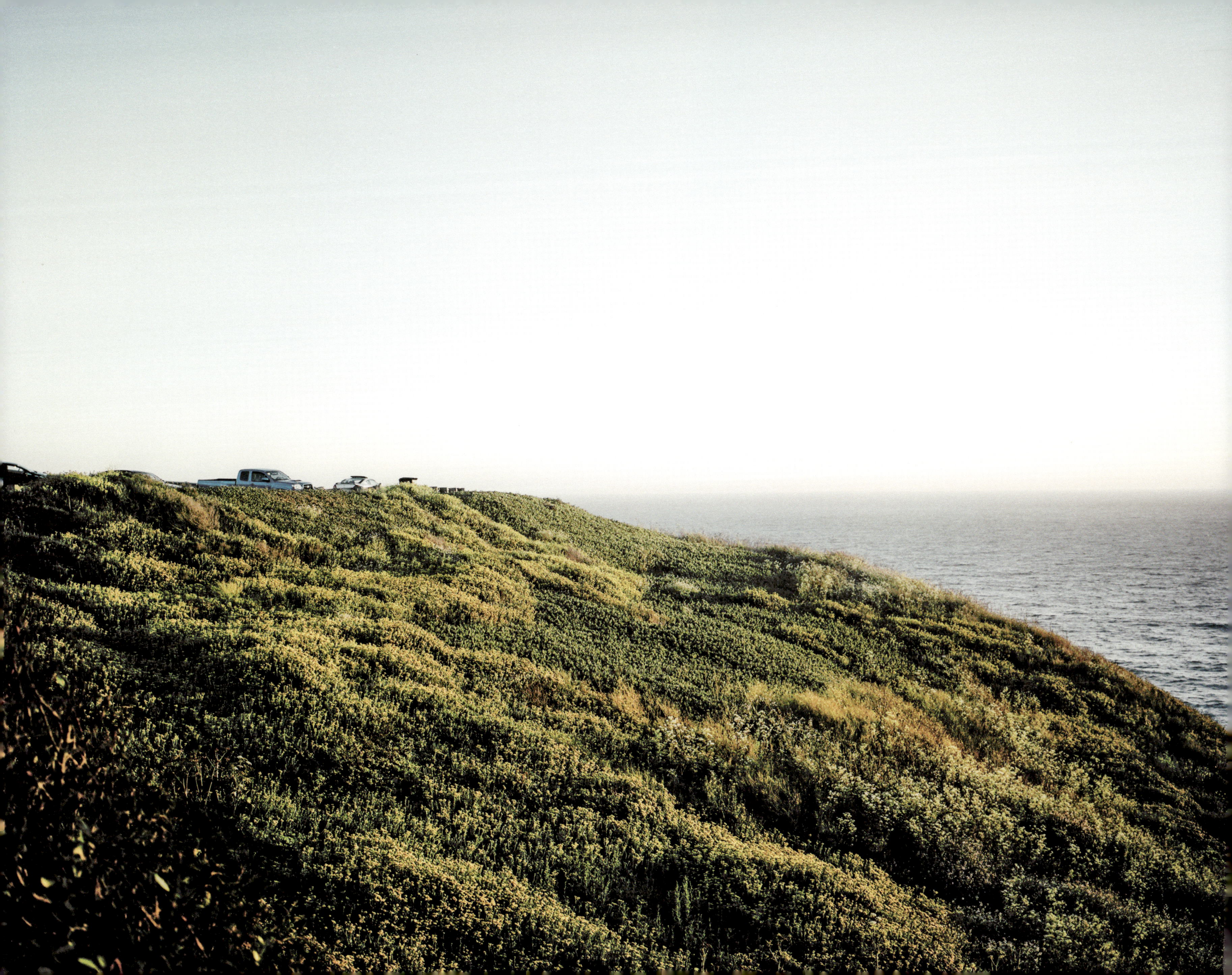

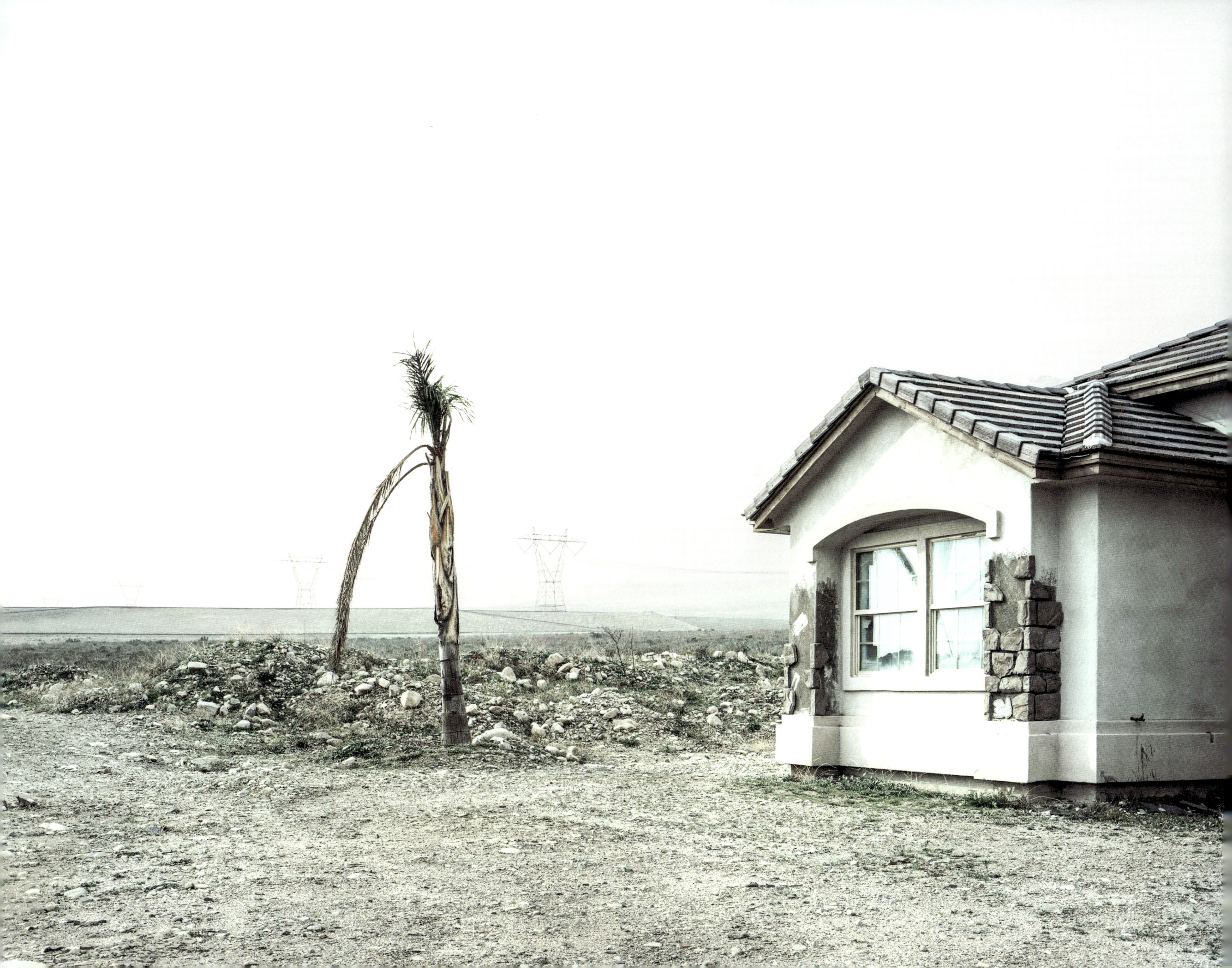

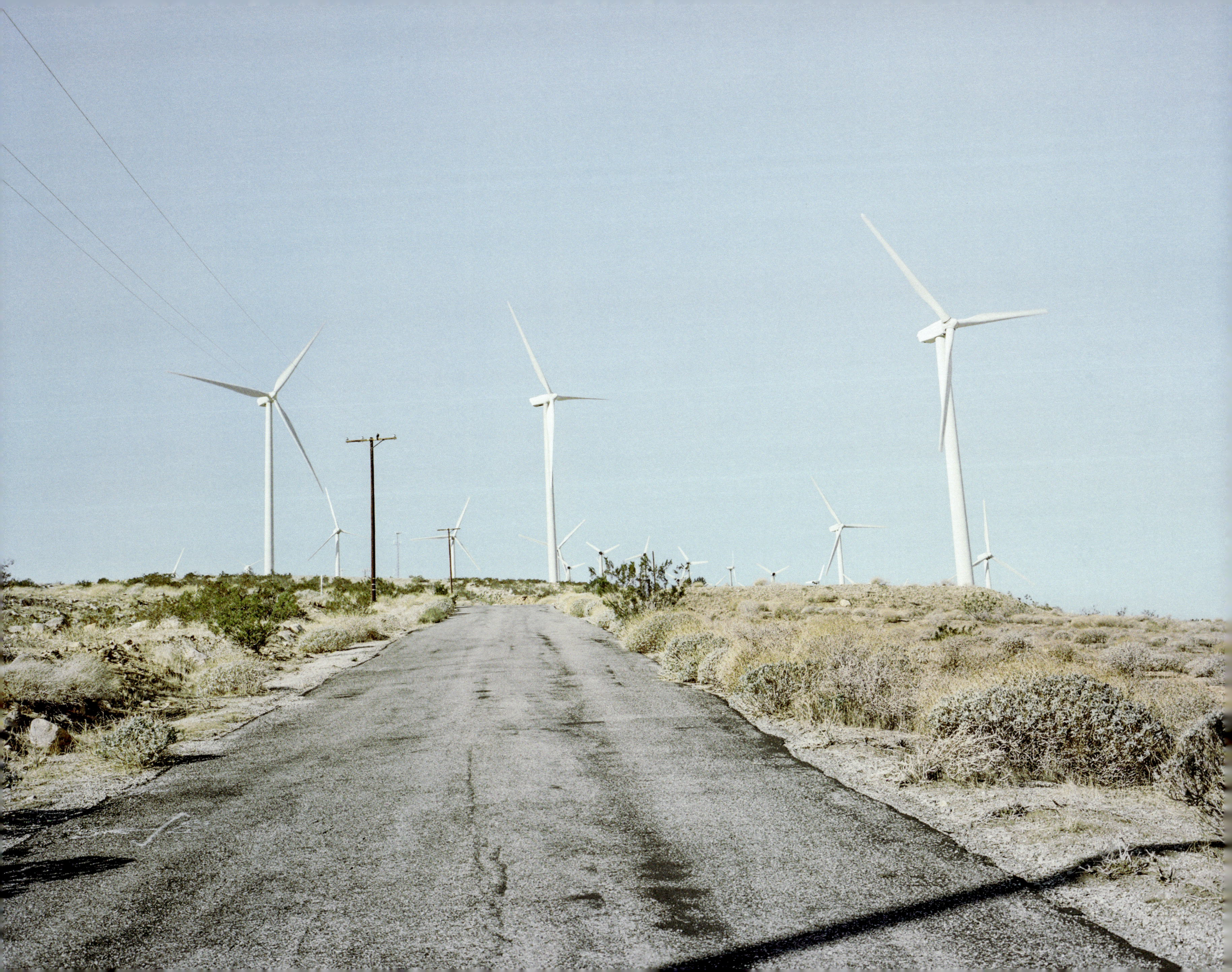

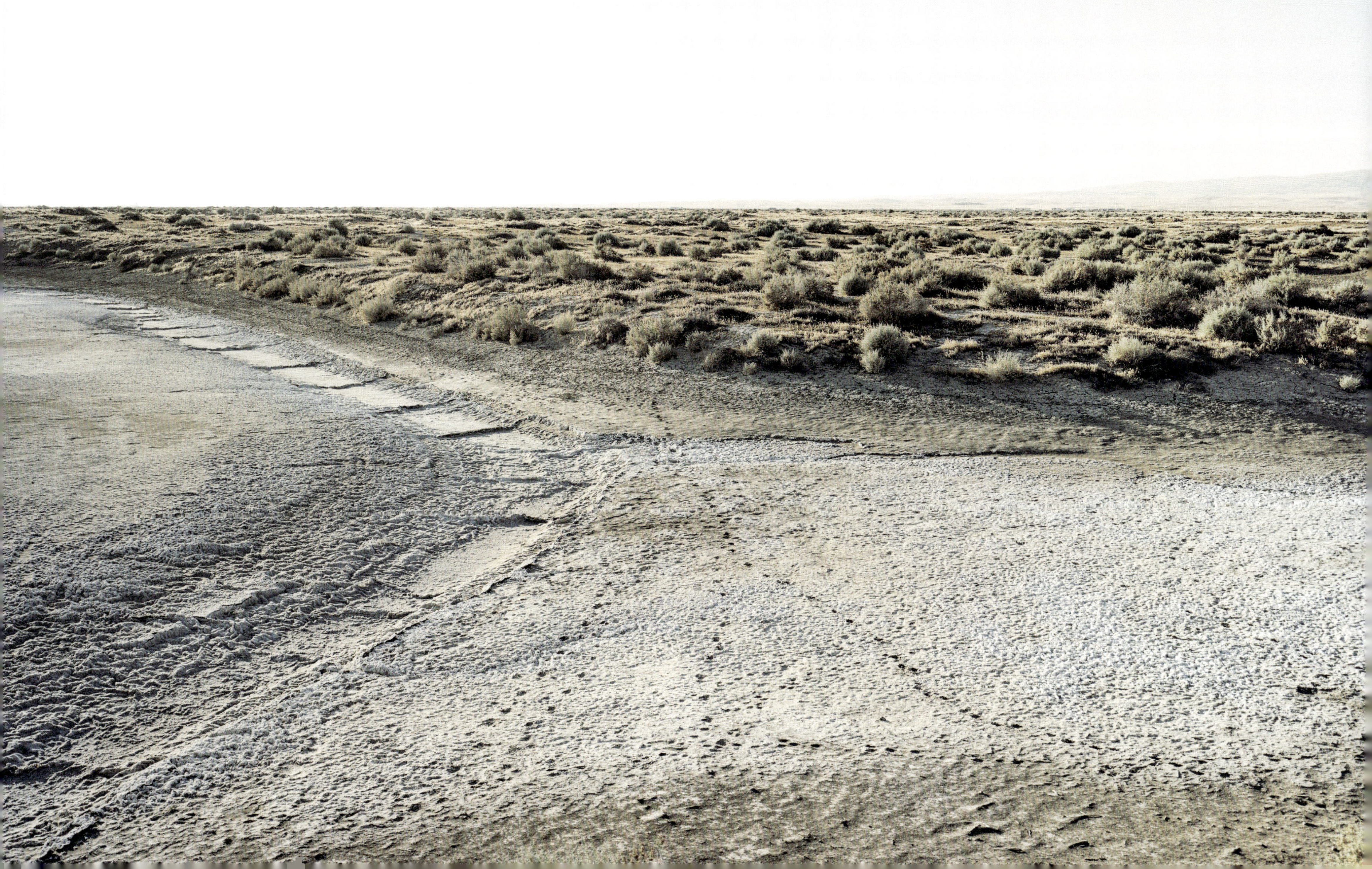

PLATE 6 | FAULT 15

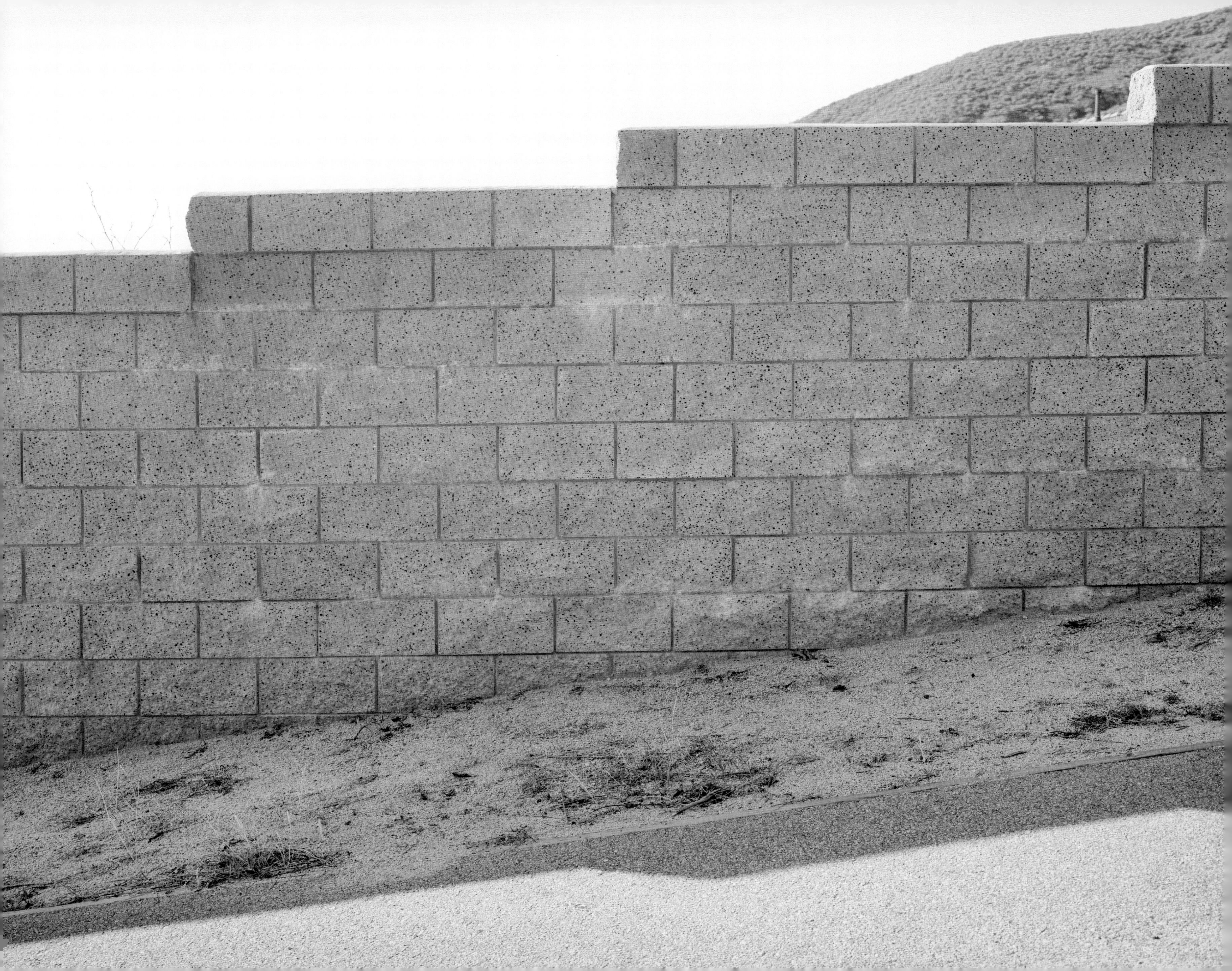

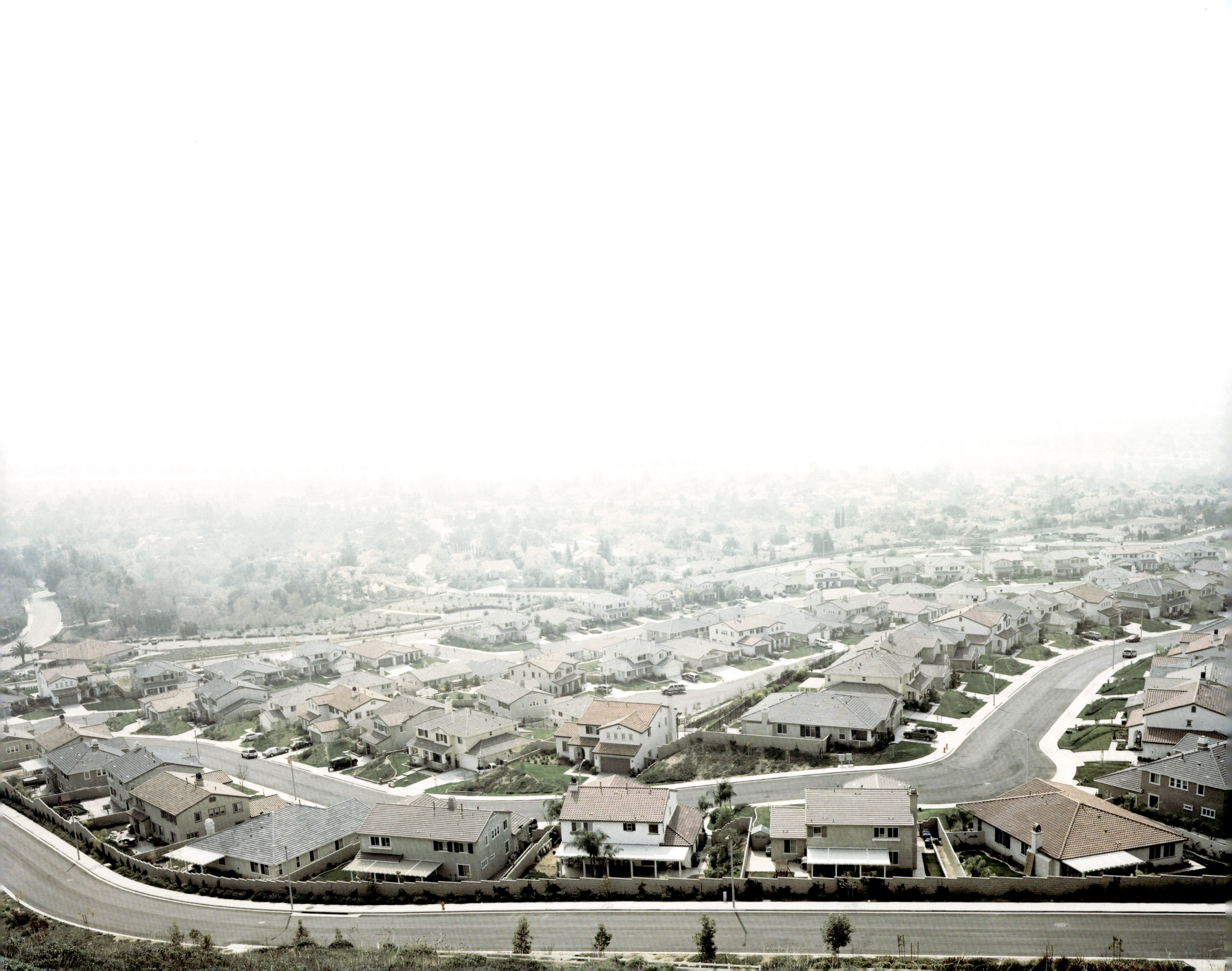

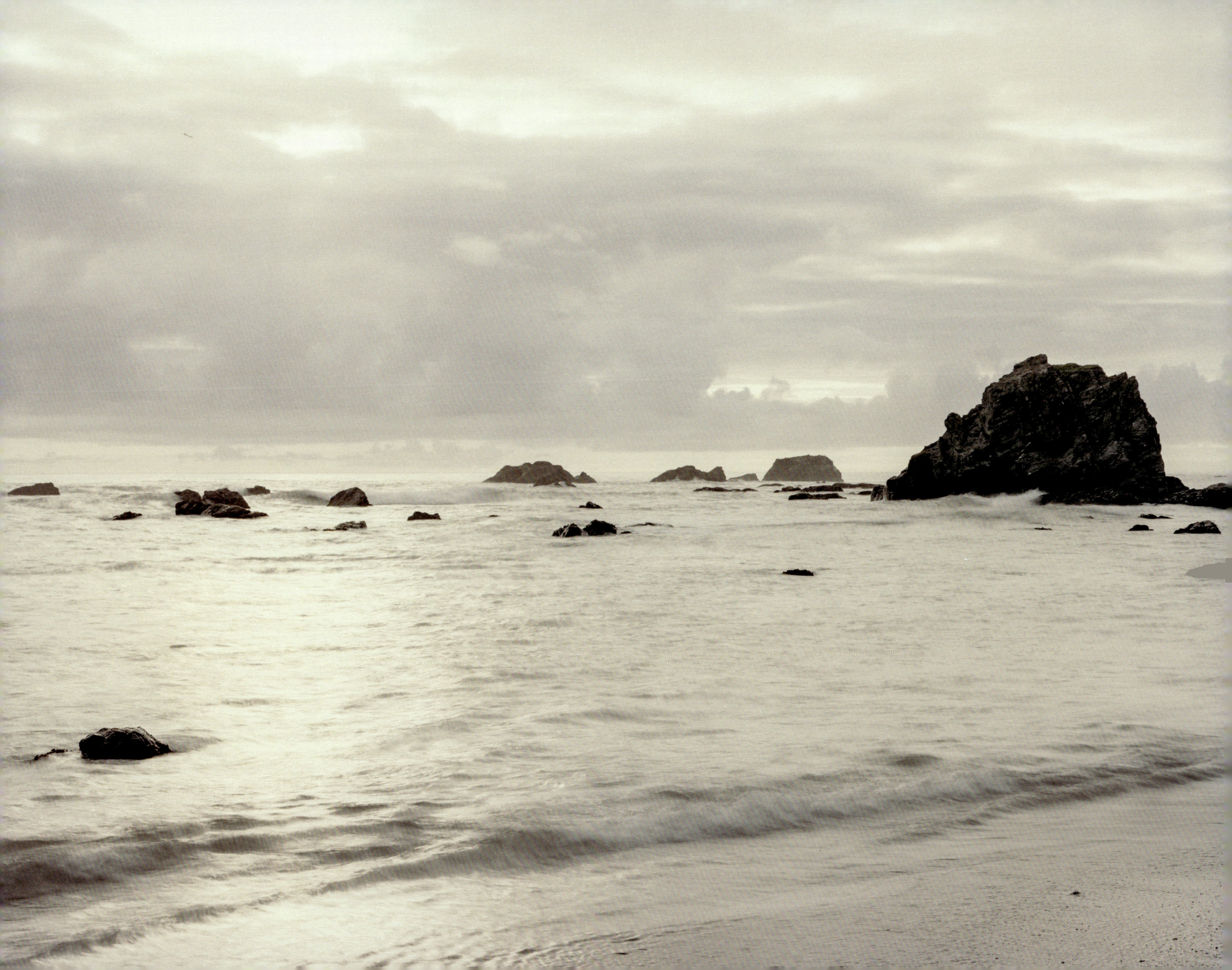

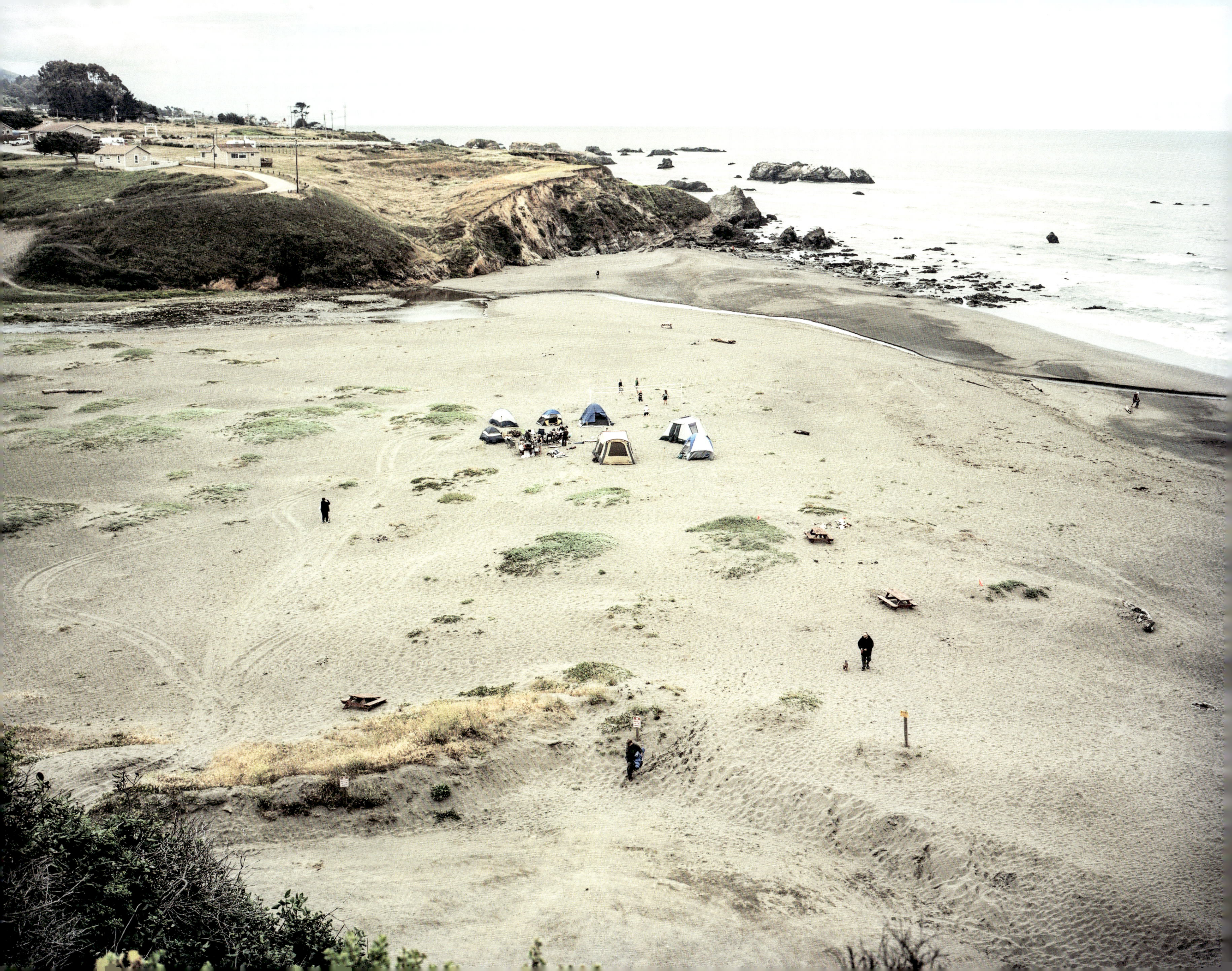

PLATE 10 | FAULT 37

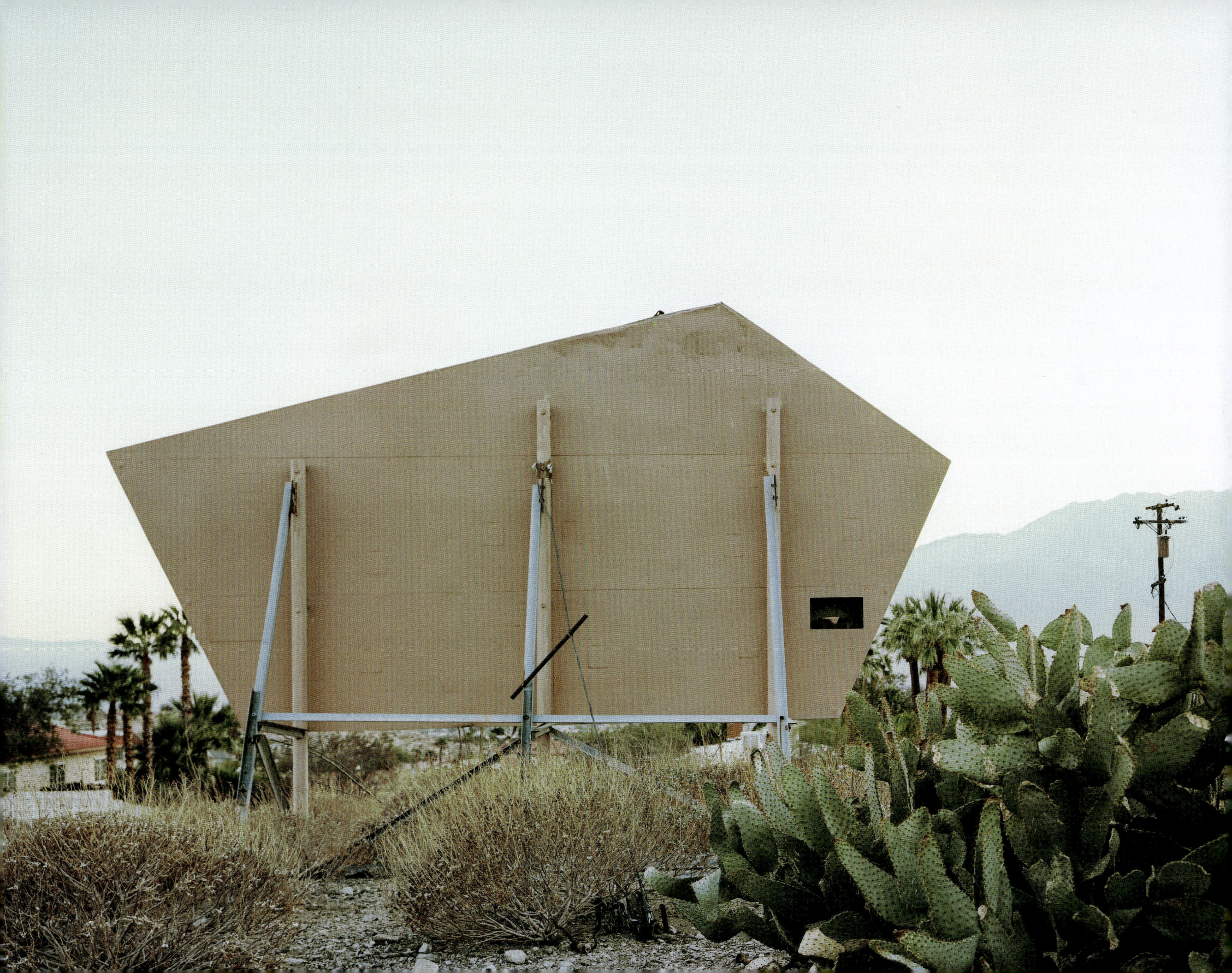

PLATE 11 | FAULT 30

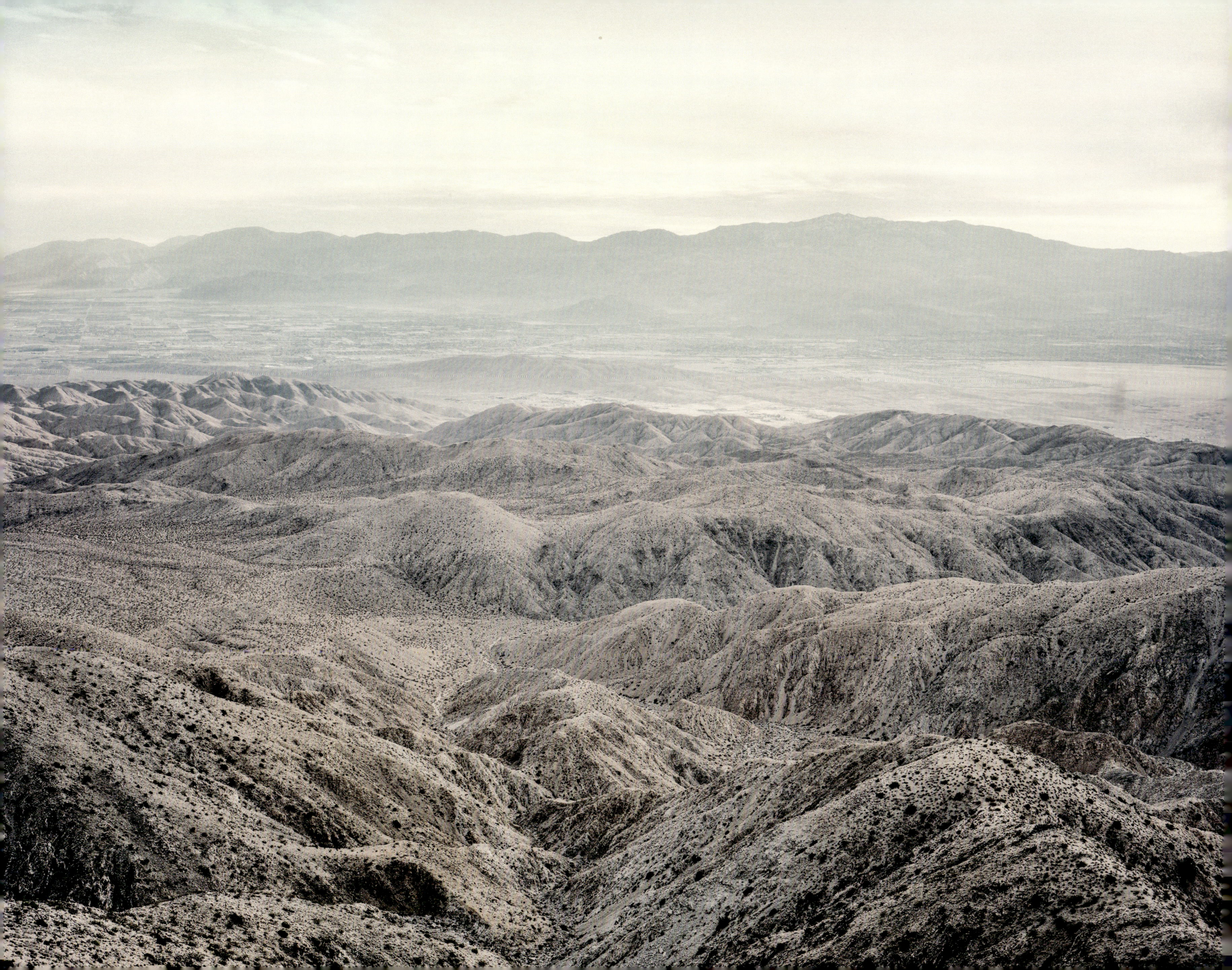

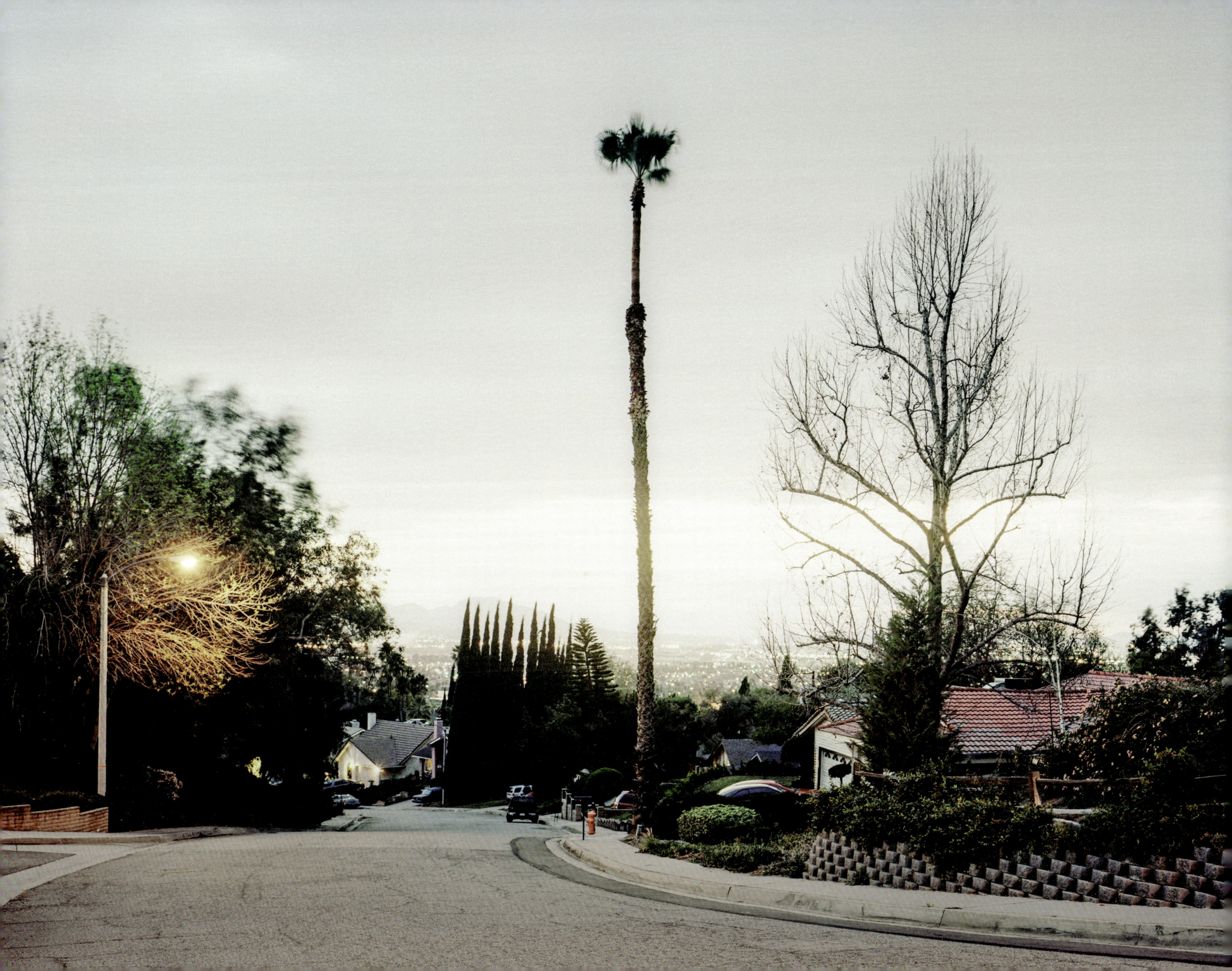

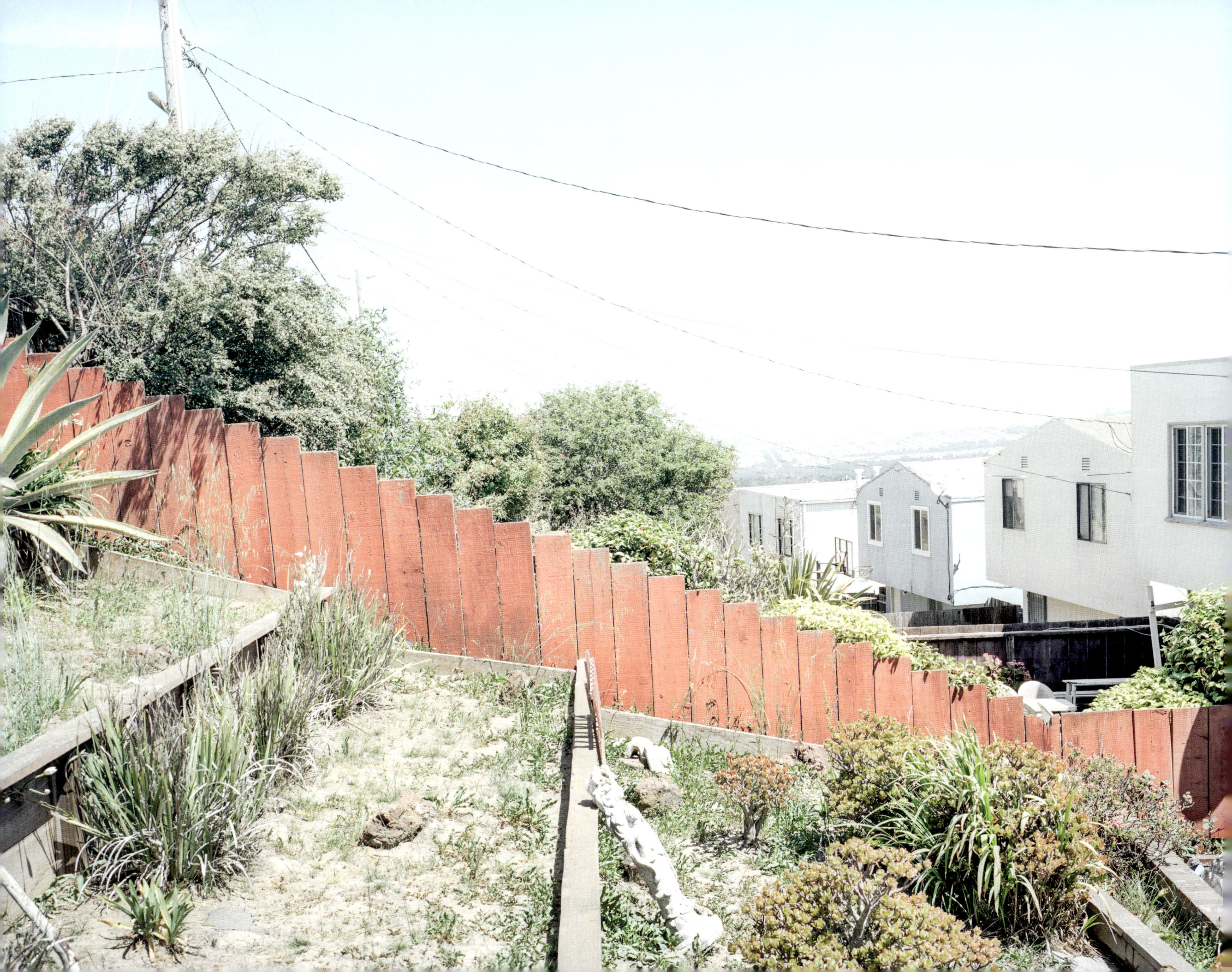

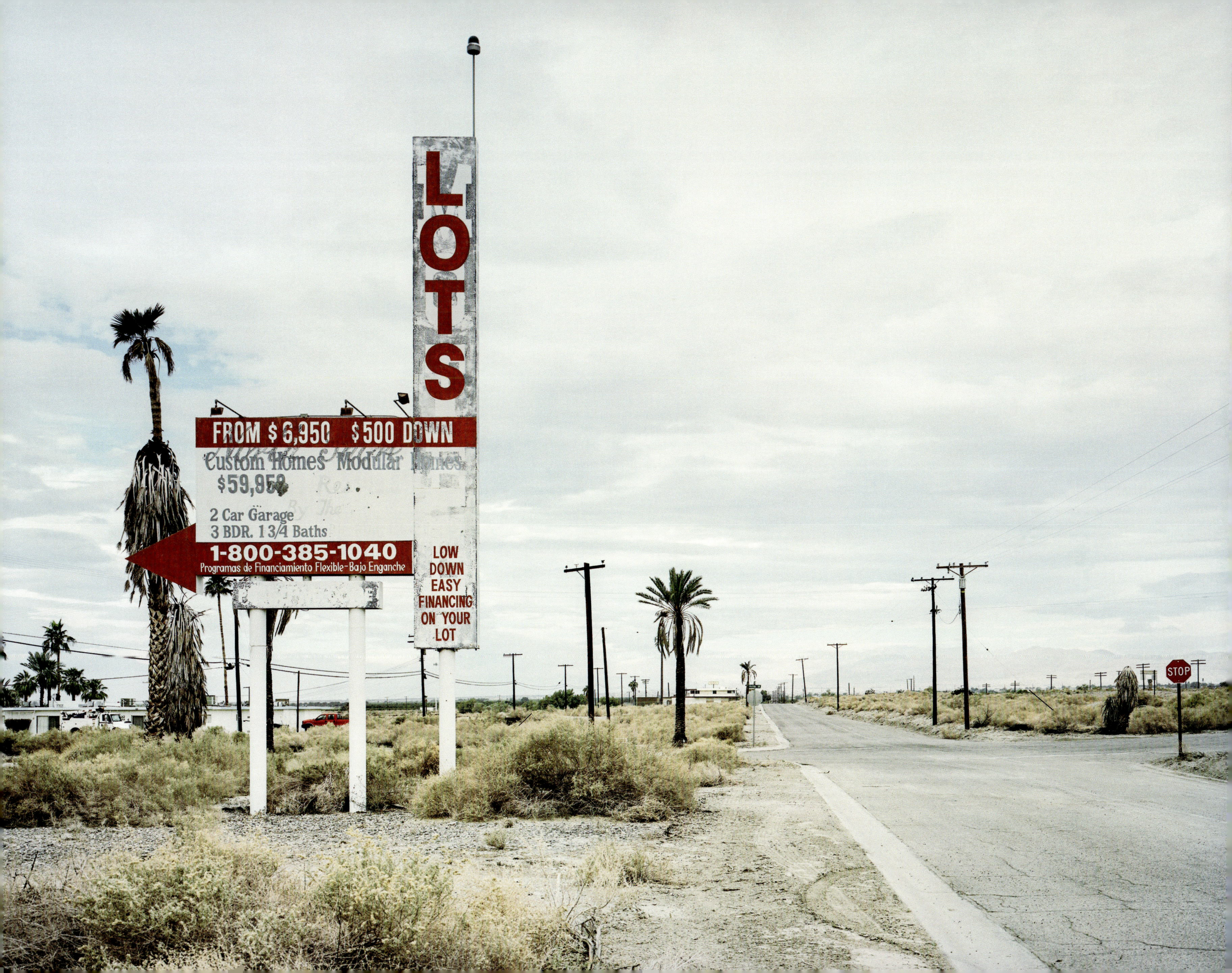

LOTS
FROM $6,950 $500 DOWN
Custom Homes Modular Homes
$59,950
2 Car Garage
3 BDR. 1 3/4 Baths
1-800-385-1040
Programas de Financiamiento Flexible-Bajo Enganche
LOW
DOWN
EASY
FINANCING
ON YOUR
LOT
STOP

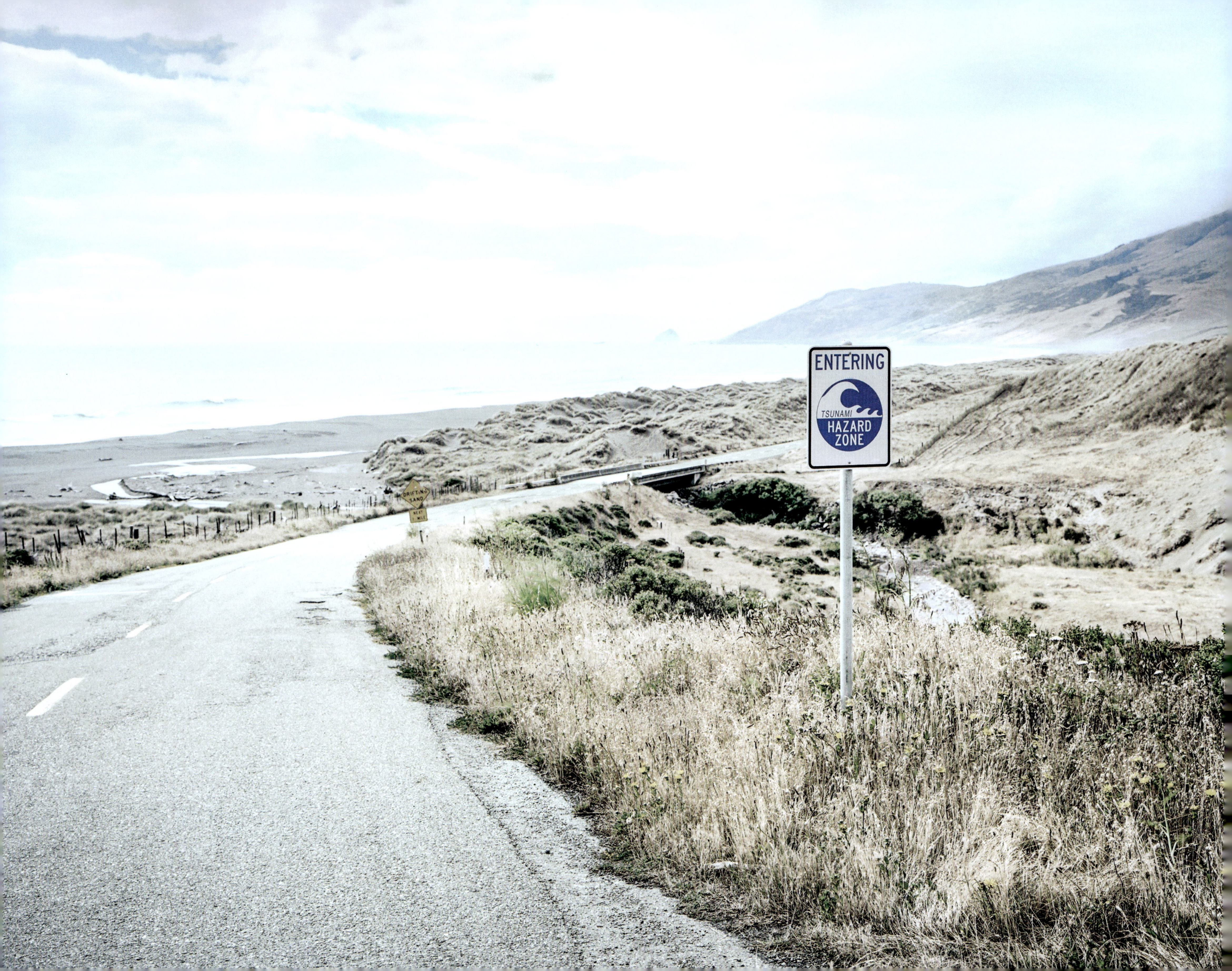
ENTERING
TSUNAMI
HAZARD
ZONE
DRIFTING
SAND
NEXT

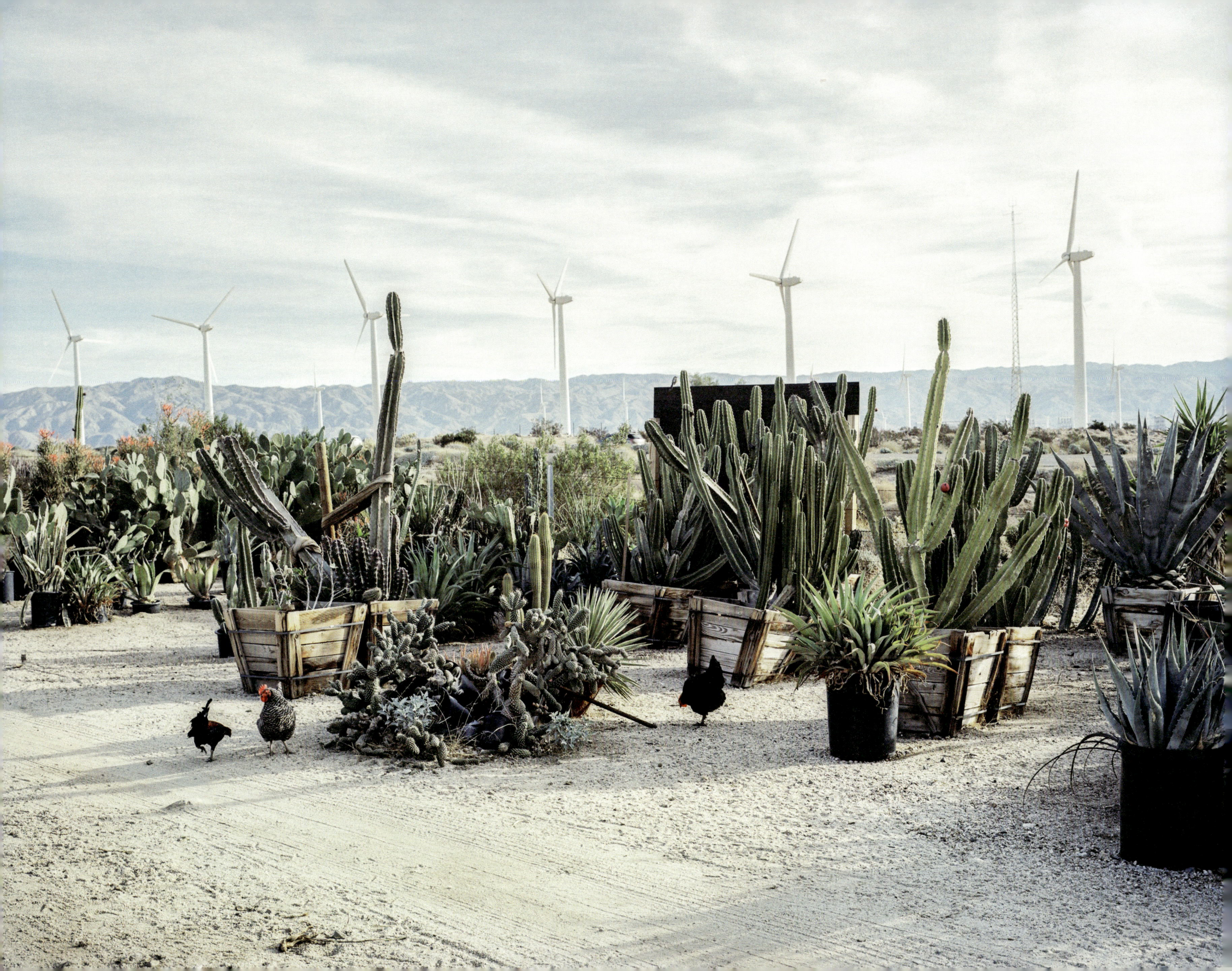

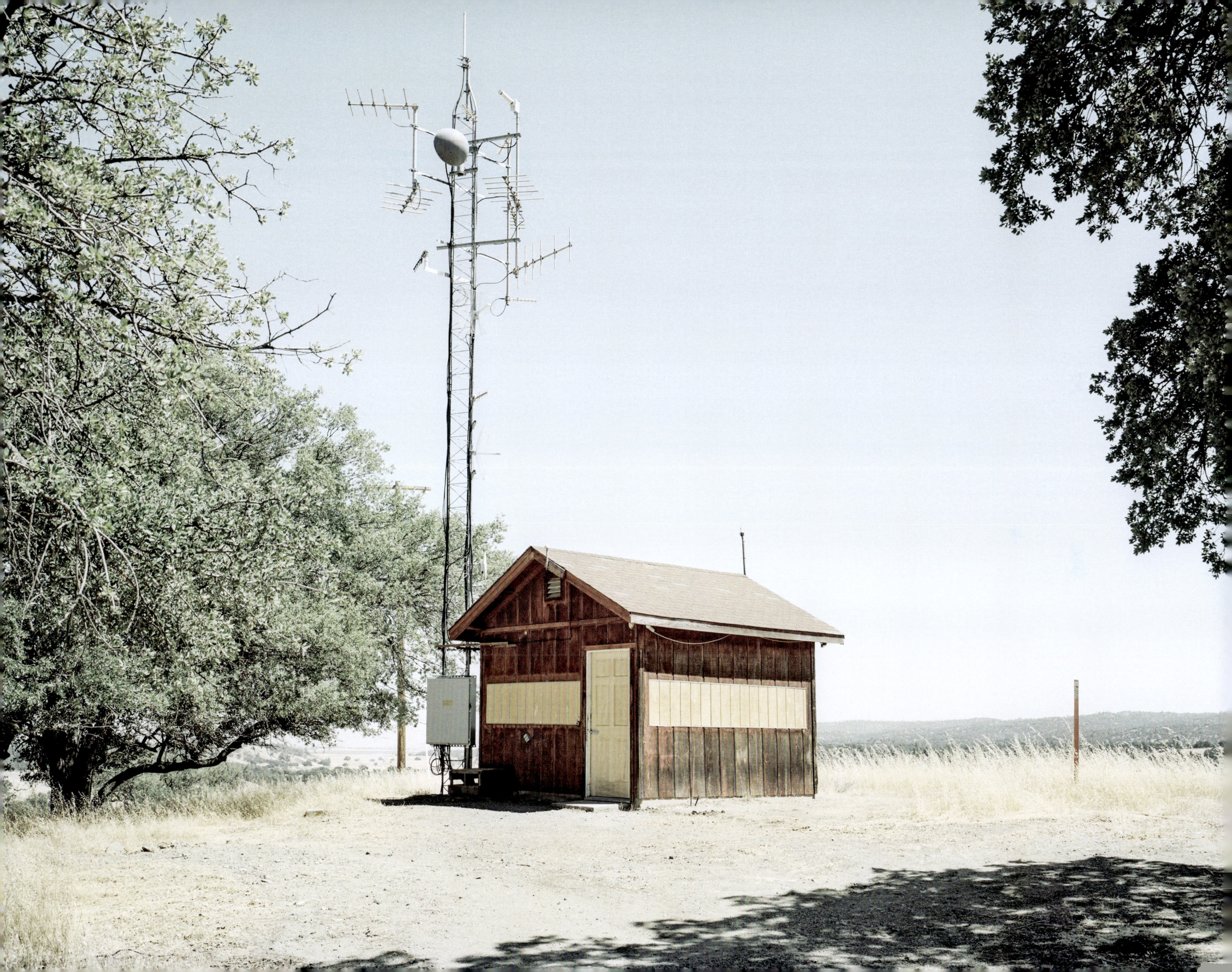

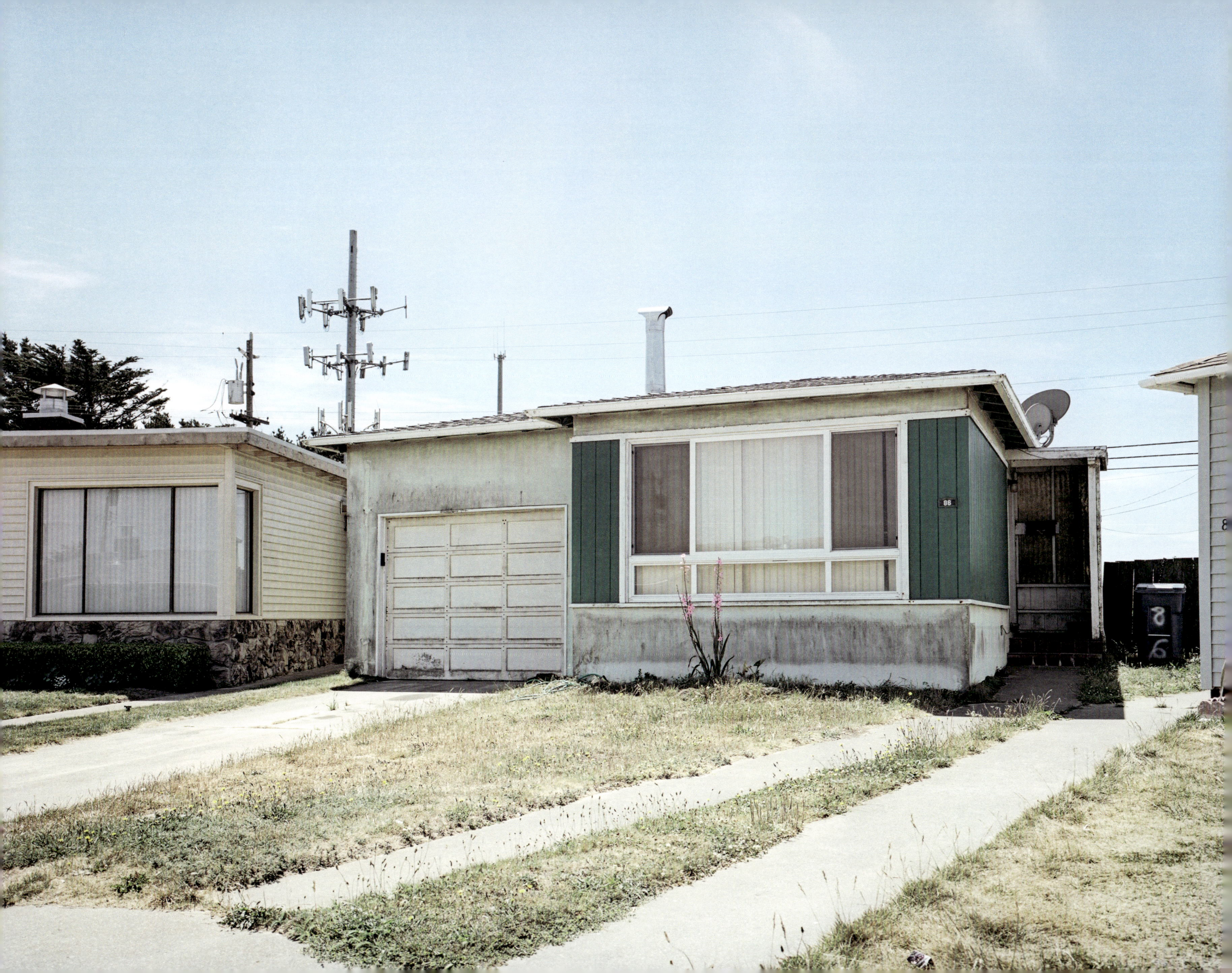

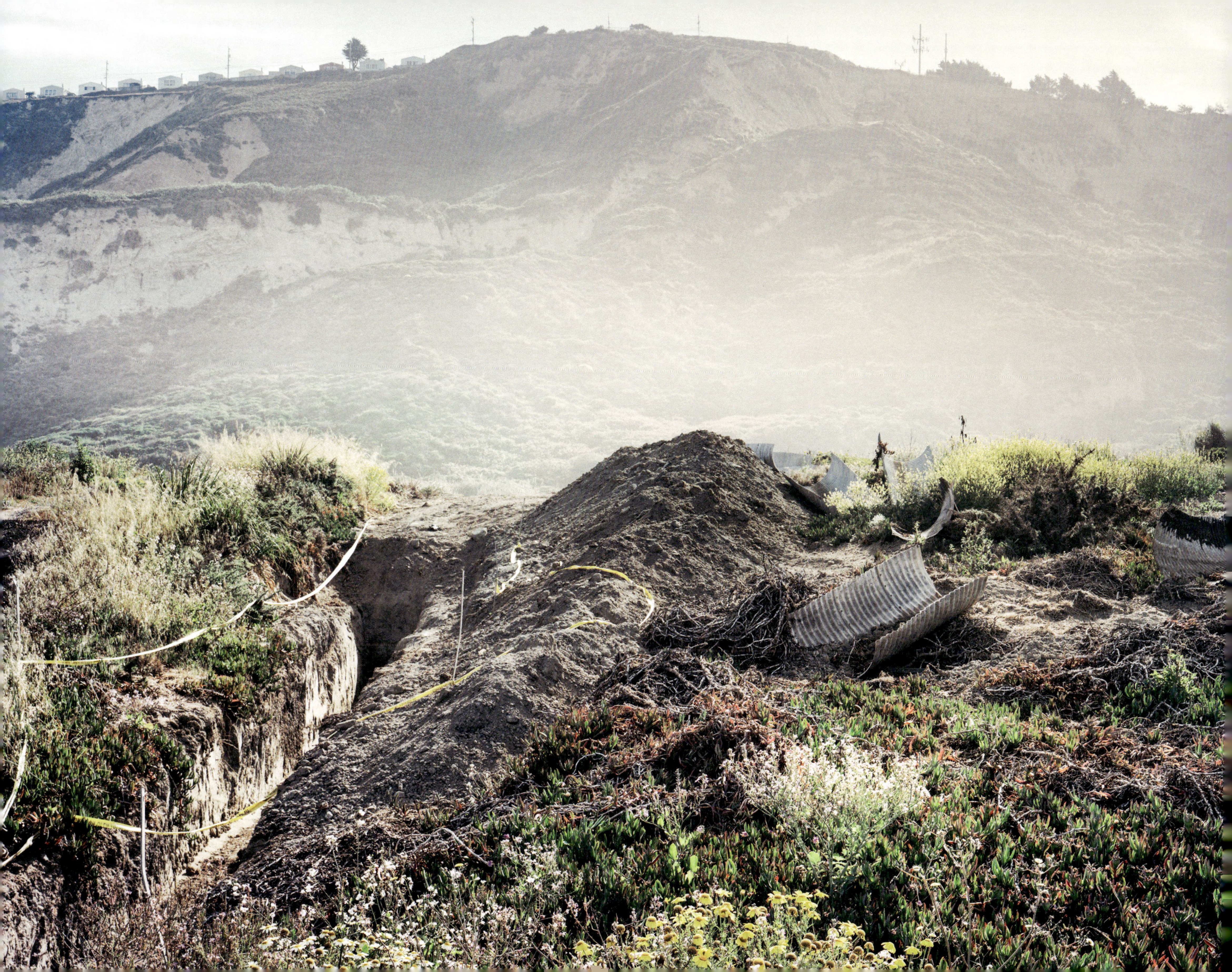

PLATE 20 | FAULT 43

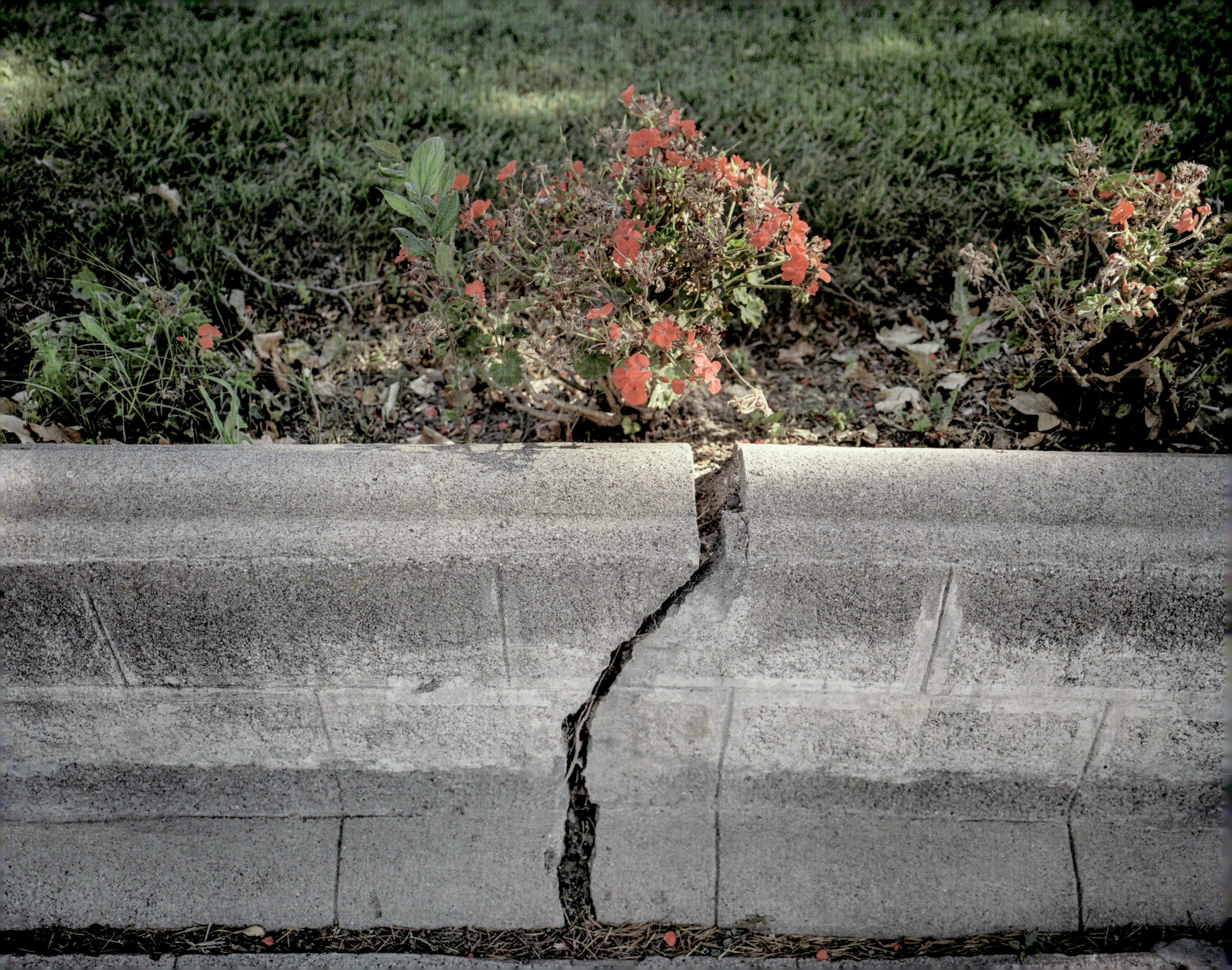

PLATE 21 | FAULT 19

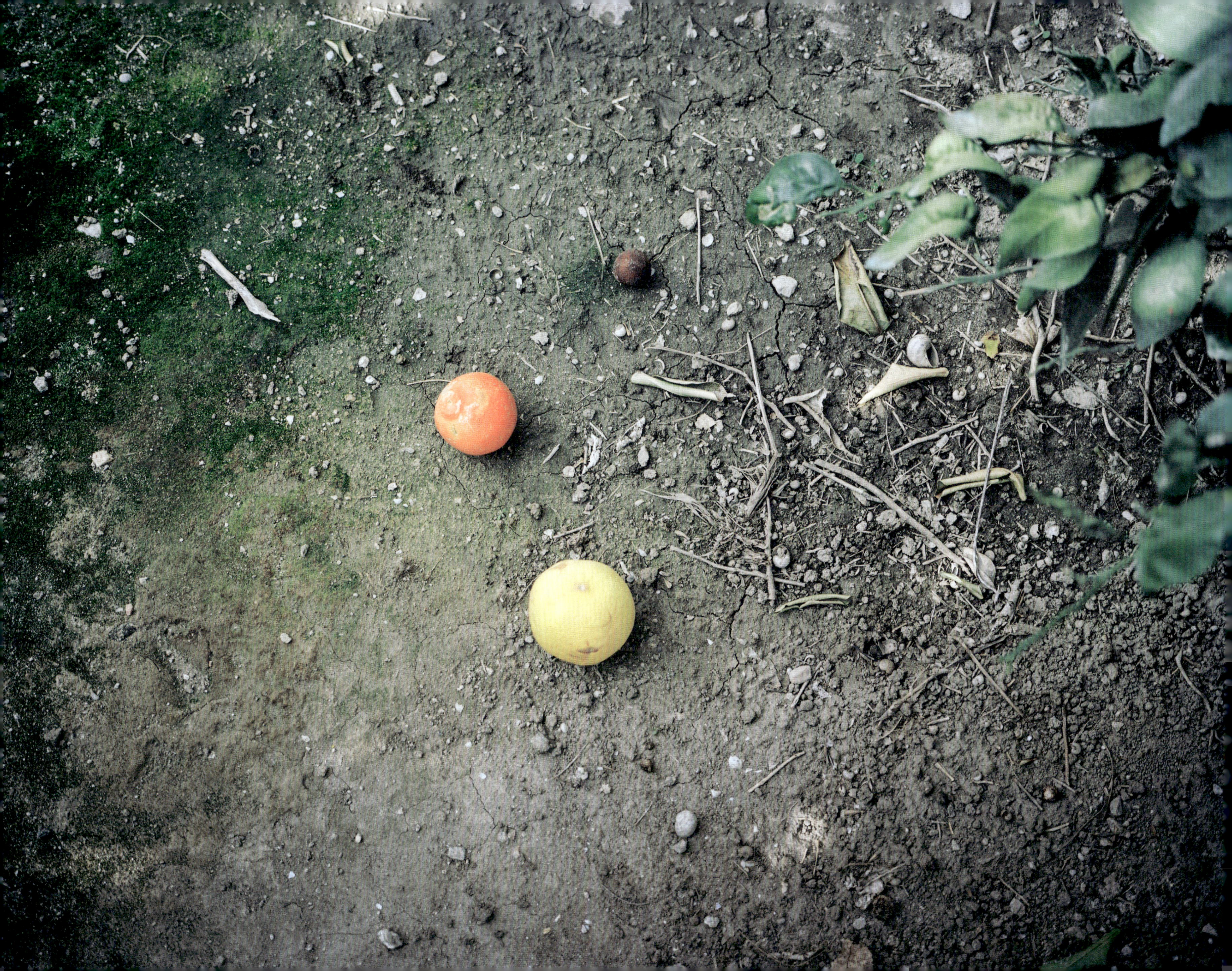

PLATE 22 | FAULT 36

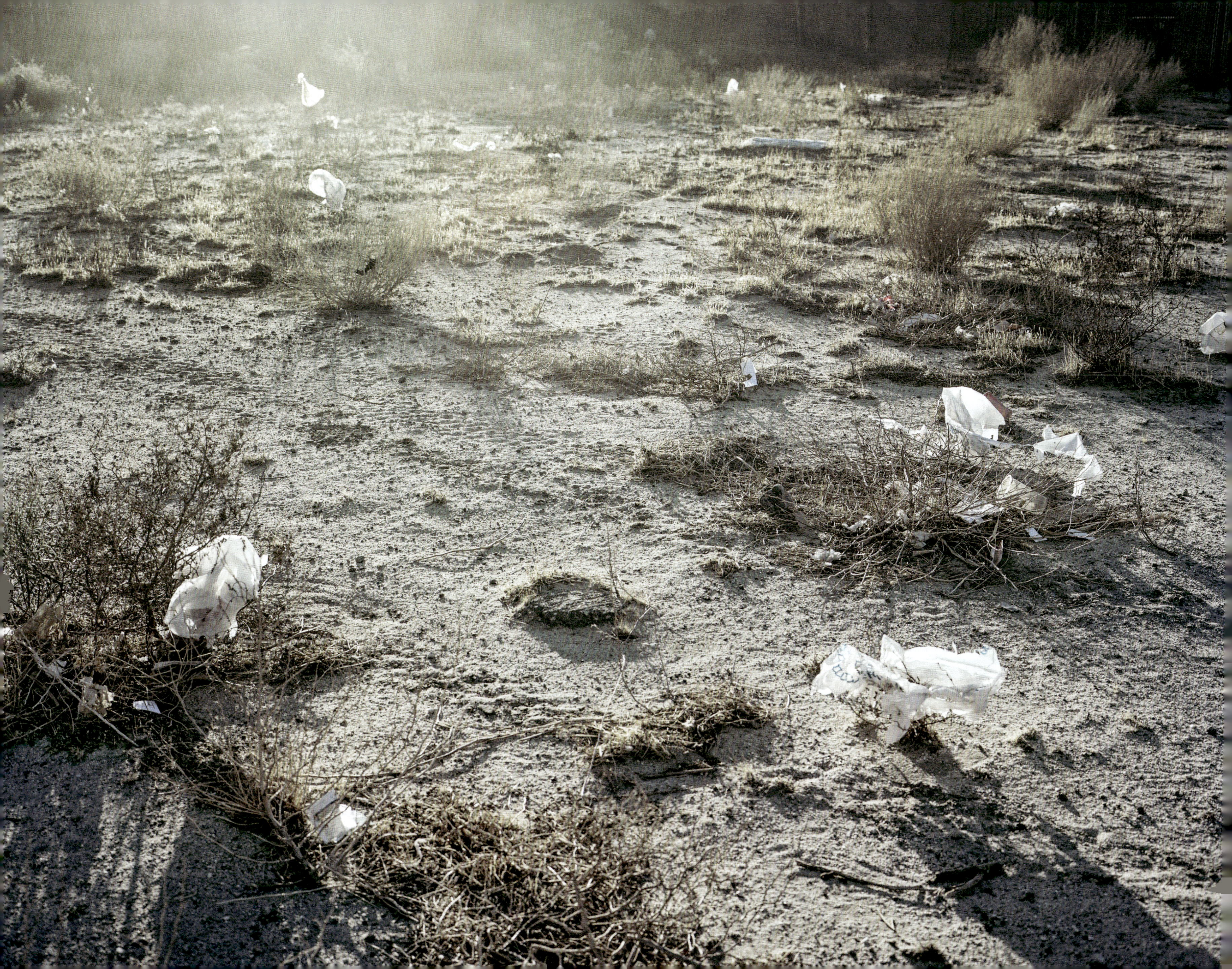

PLATE 23 | FAULT 3

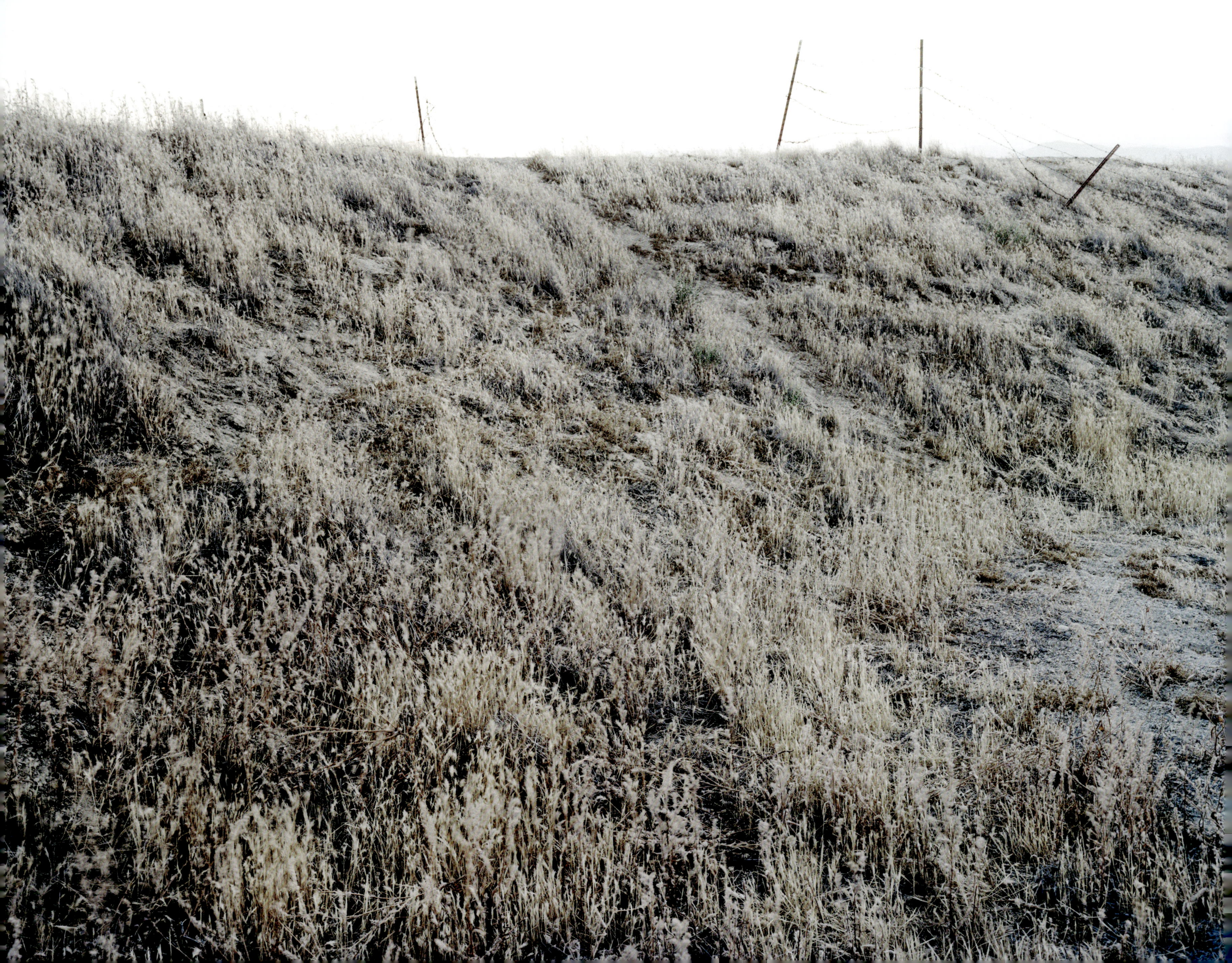

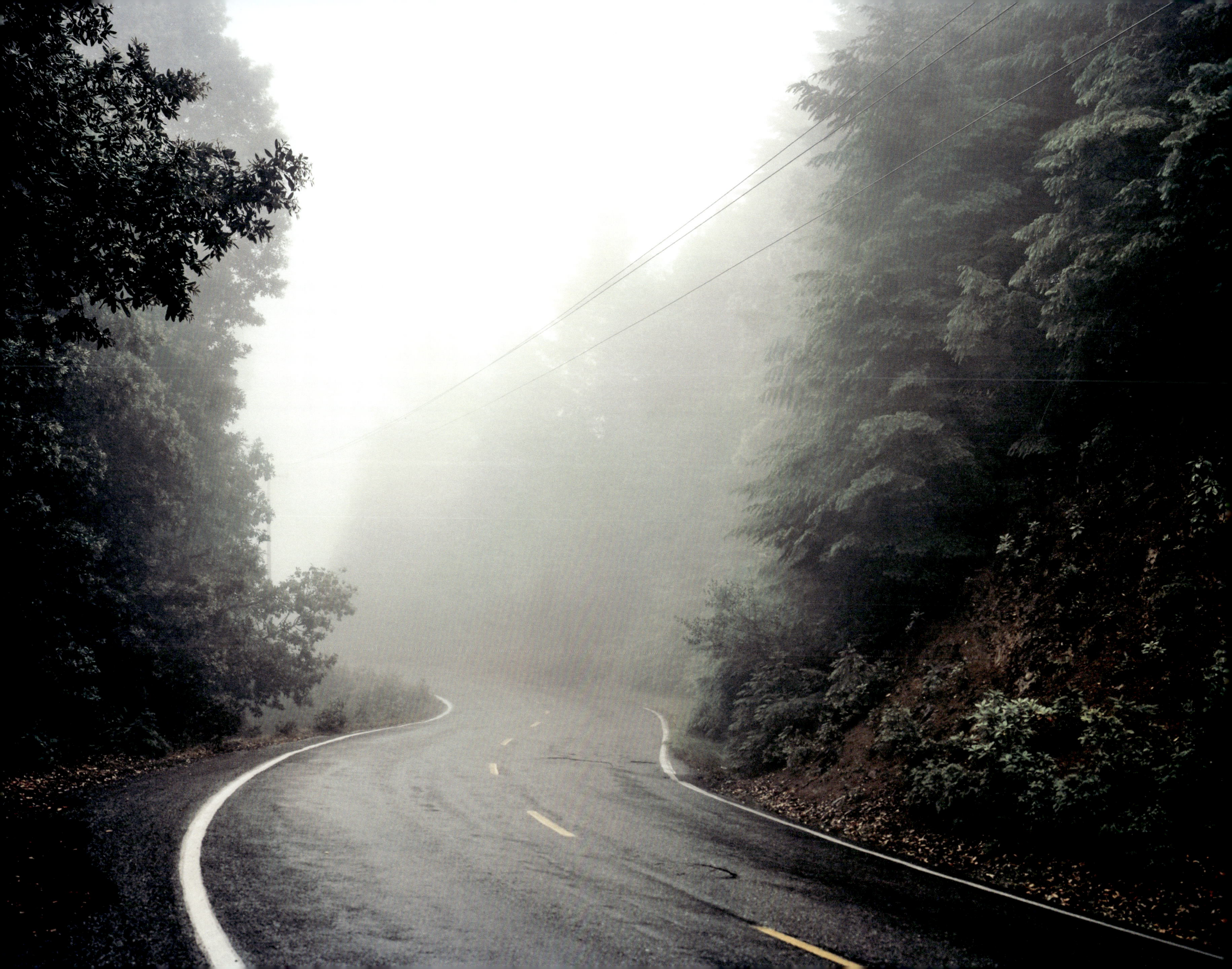

RADIUS BOOKS

227 E. Palace Ave. Suite W Santa Fe, NM 87501 t: (505) 983–4068 www.radiusbooks.org

Available through

D.A.P. / DISTRIBUTED ART PUBLISHERS

155 Sixth Ave., 2nd Floor, New York, NY 10013 t: (212) 627–1999 www.artbook.com

ISBN 978-1-942185-15-4

Library of Congress Cataloging-in-Publication Data available from the publisher upon request.

DESIGN: David Chickey

PROOFREADING: Miranda Pennington

PRE-PRESS: John Vokoun

Printed by Editoriale Bortolazzi-Stei, Verona, Italy

RIFT

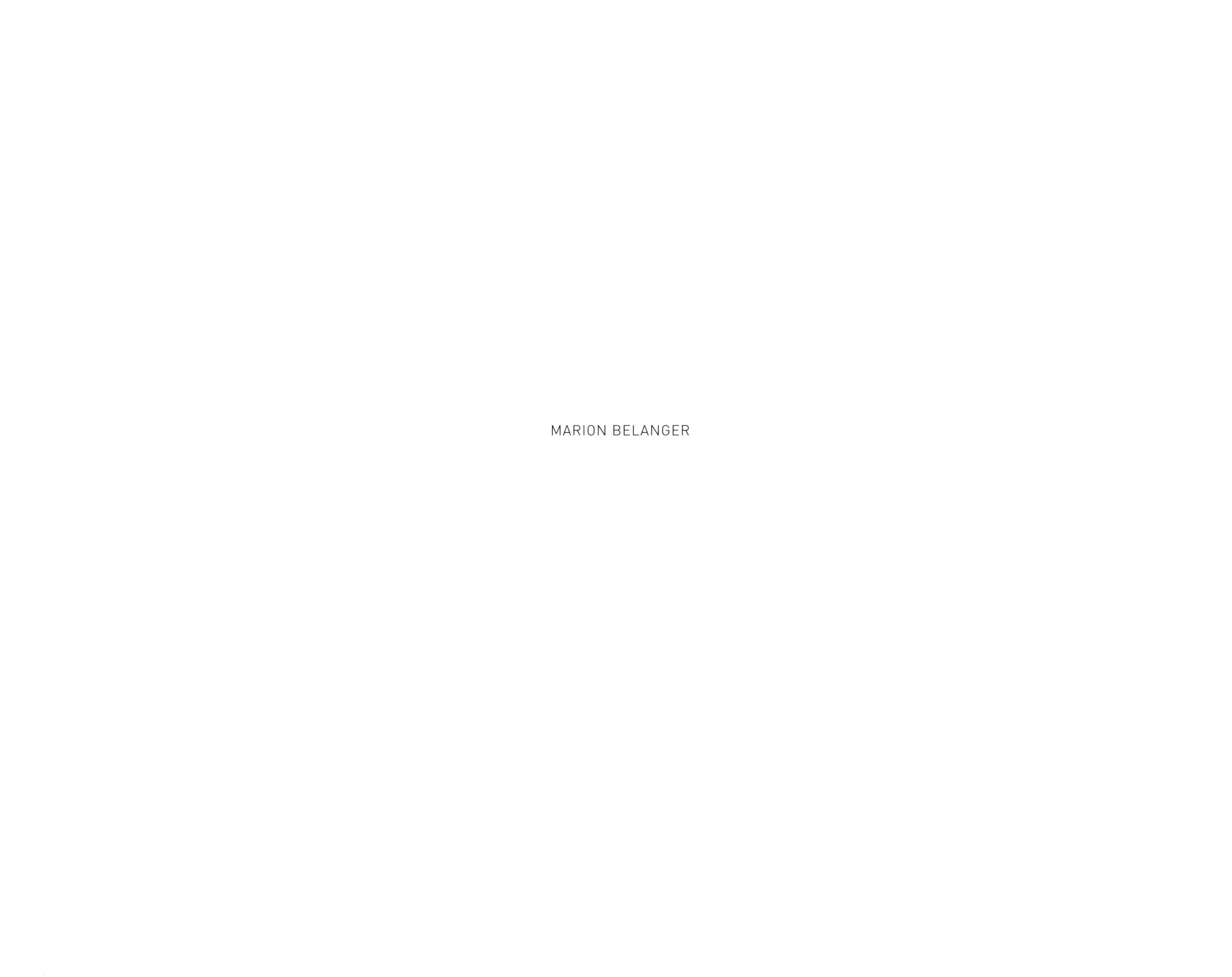

MARION BELANGER

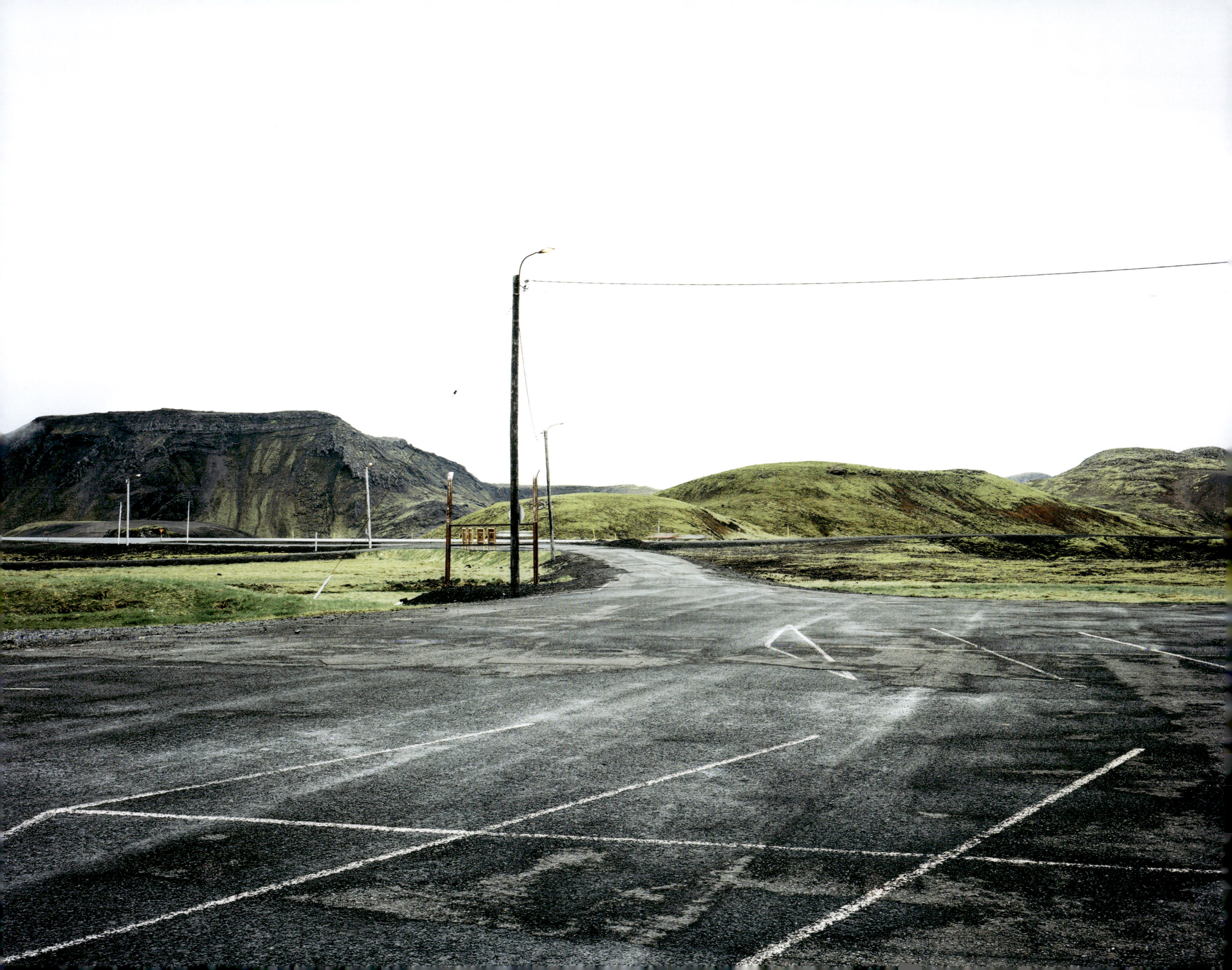

PLATE 2 | RIFT 8

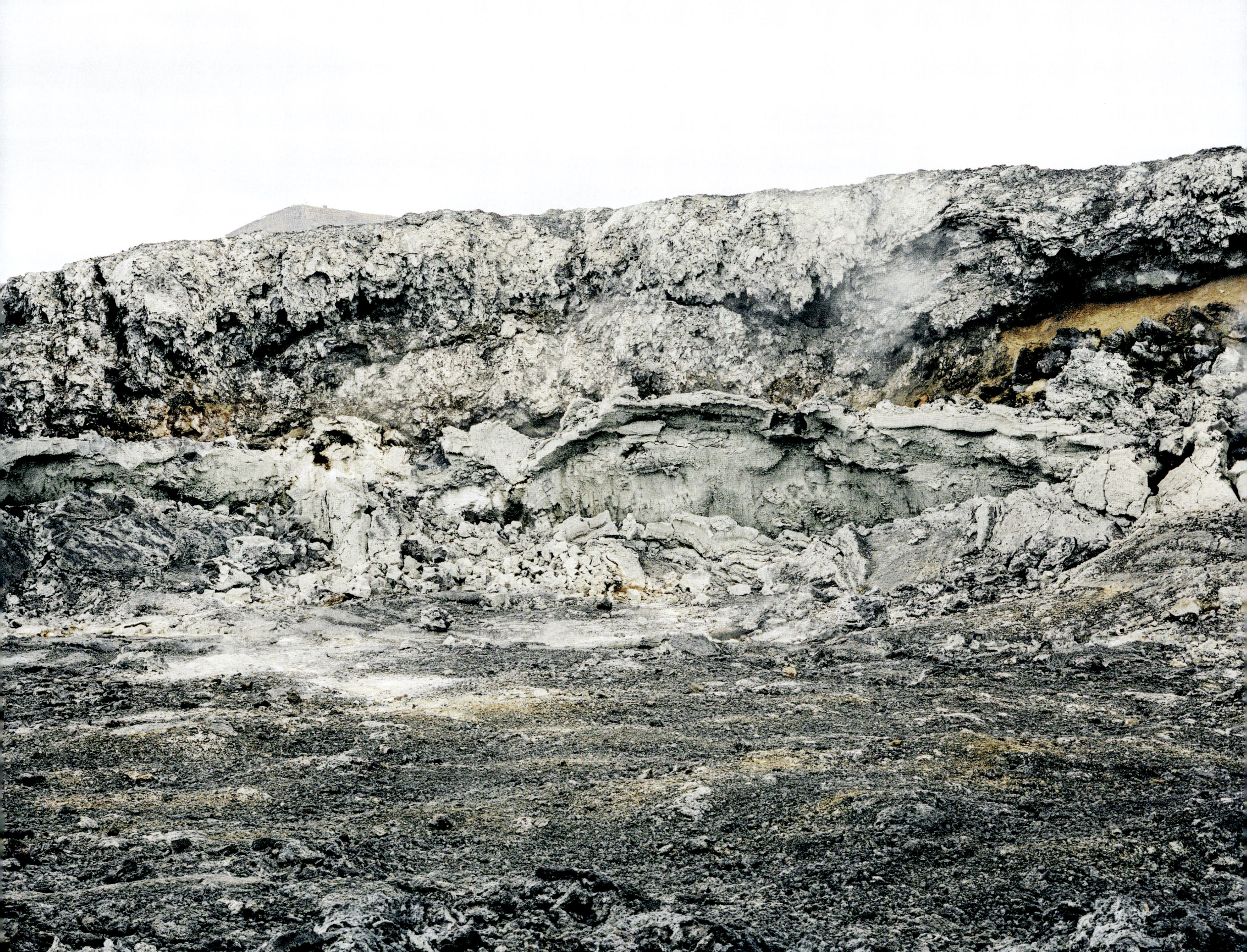

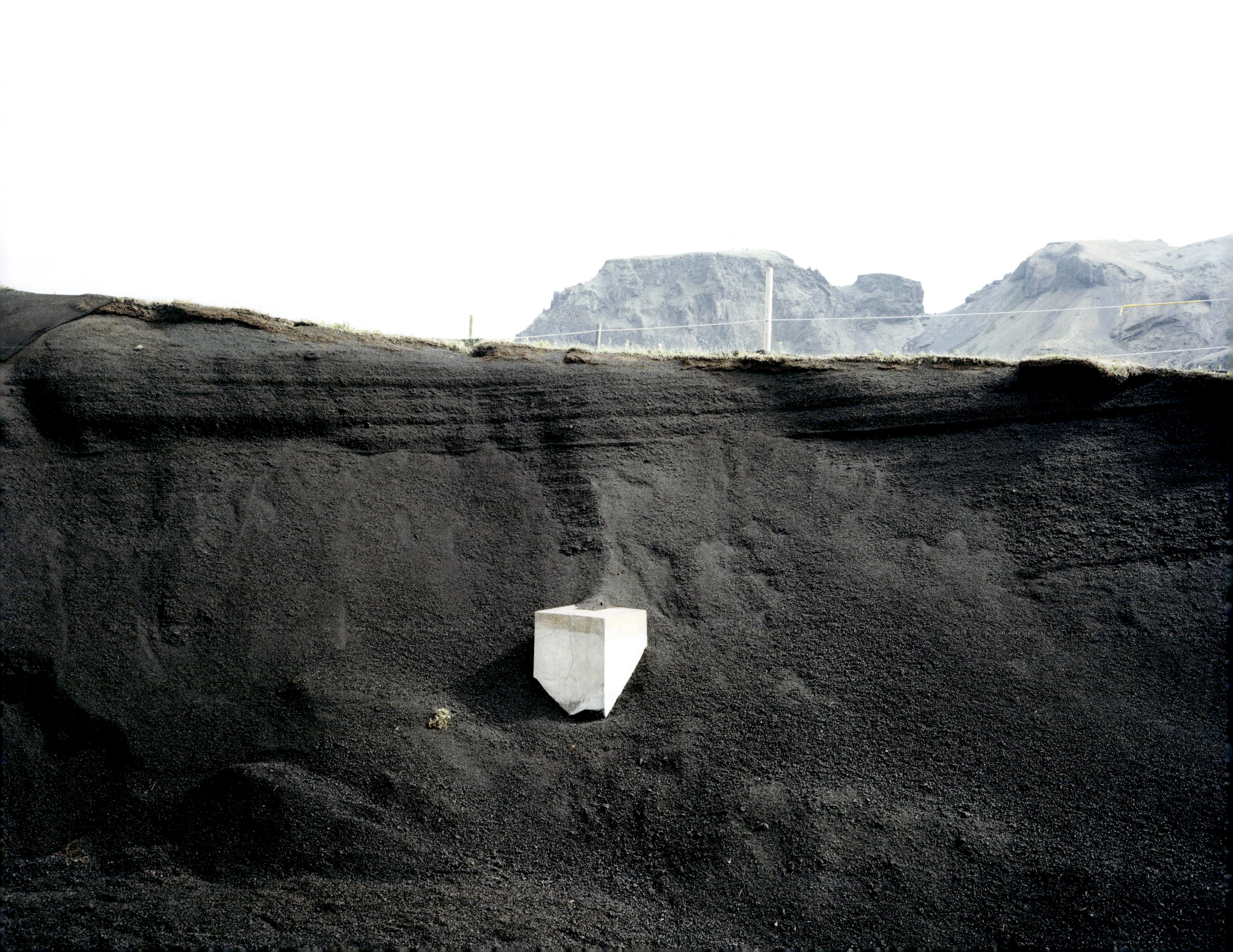

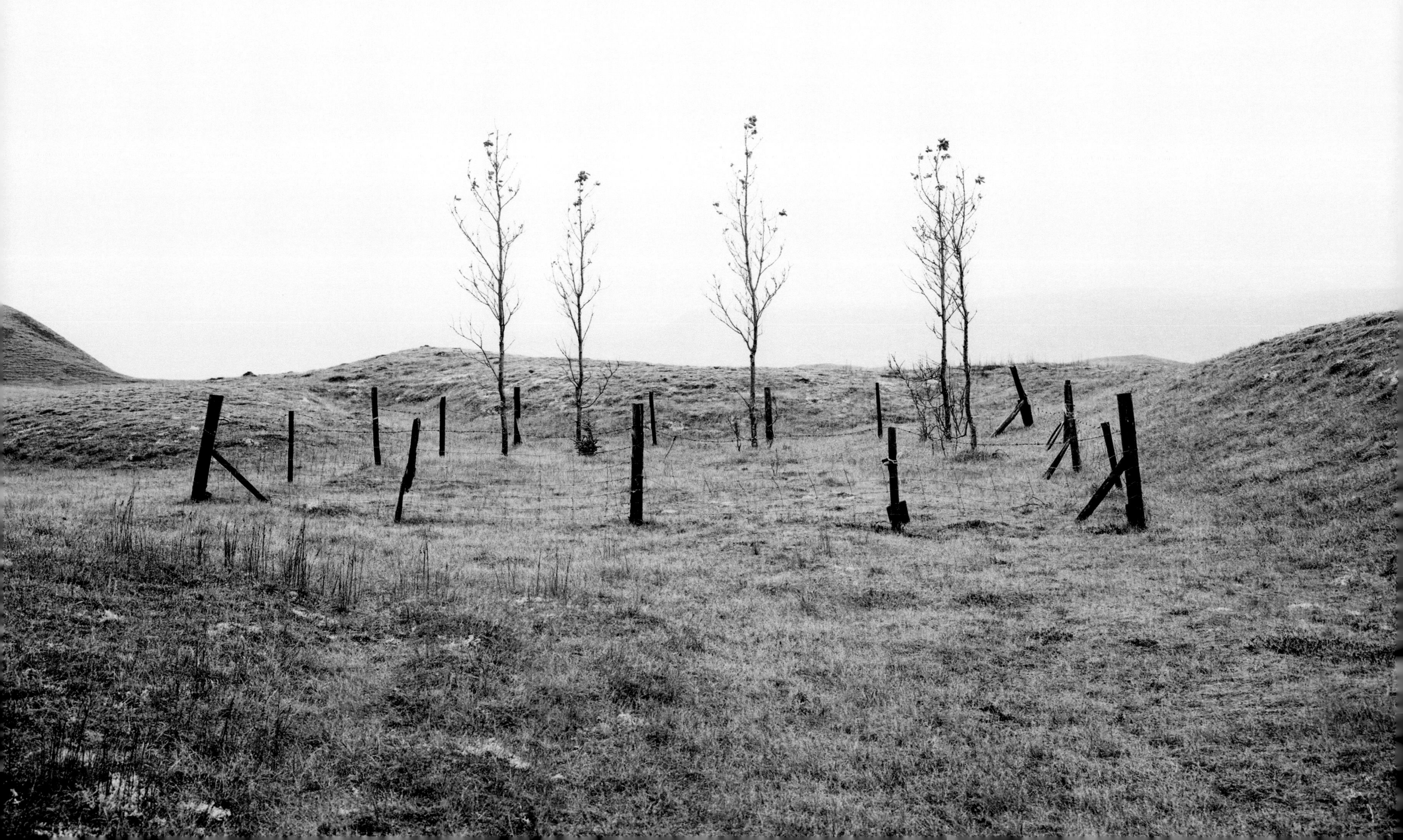

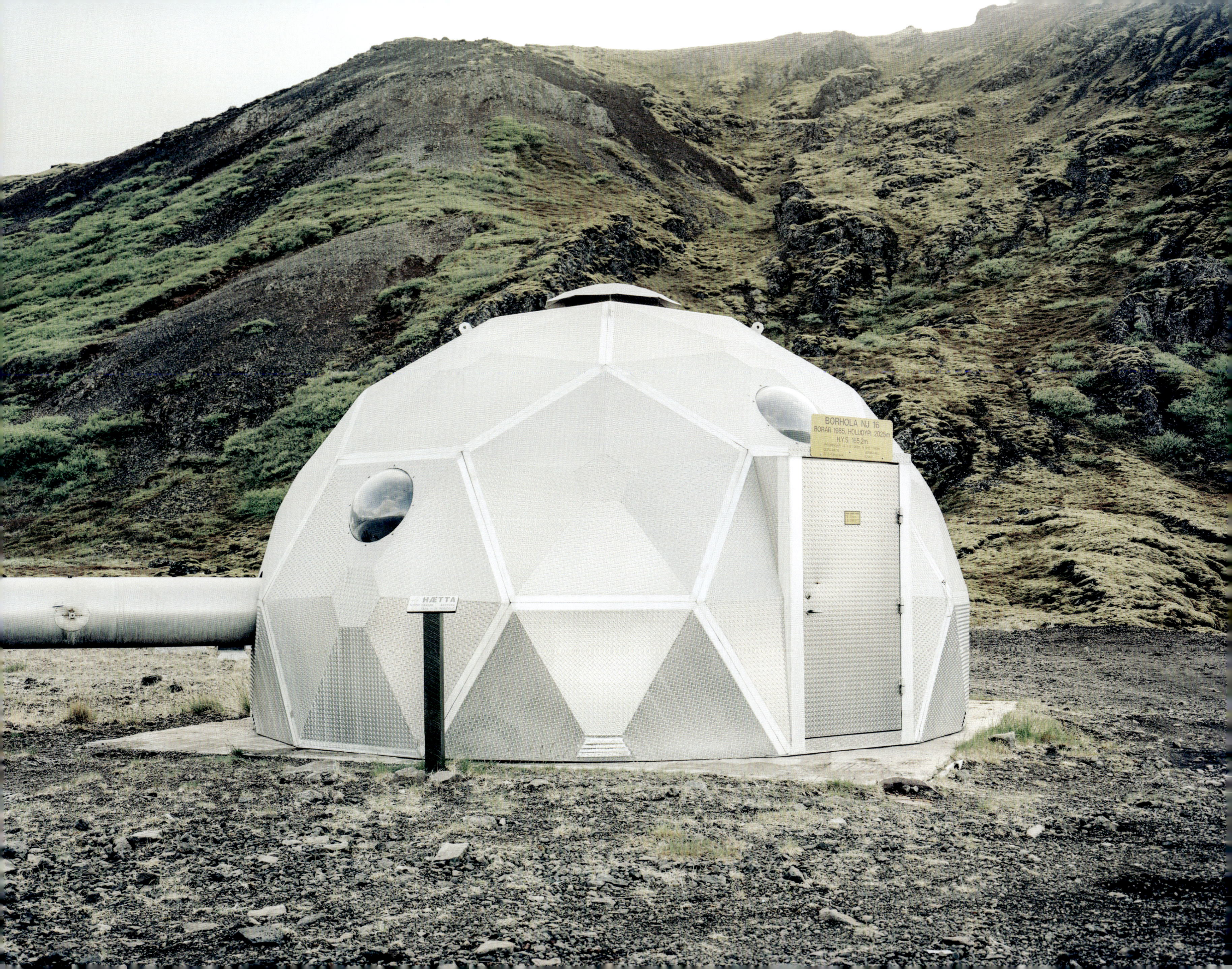

BORHOLA NJ 16
BORÁR 1985. HOLUDÝPI 2025m
H.Y.S. 165.2m
HÆTTA

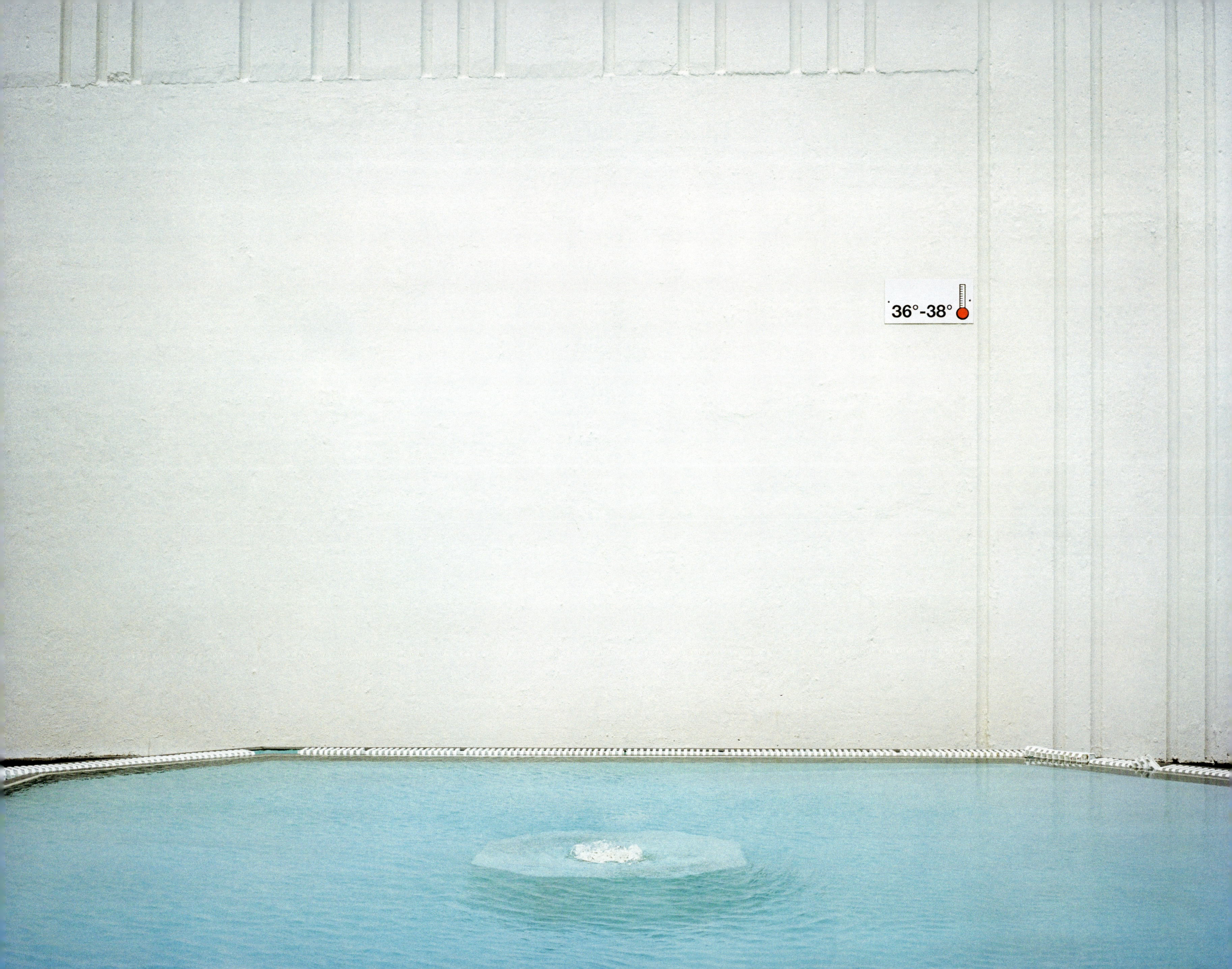
36°-38°

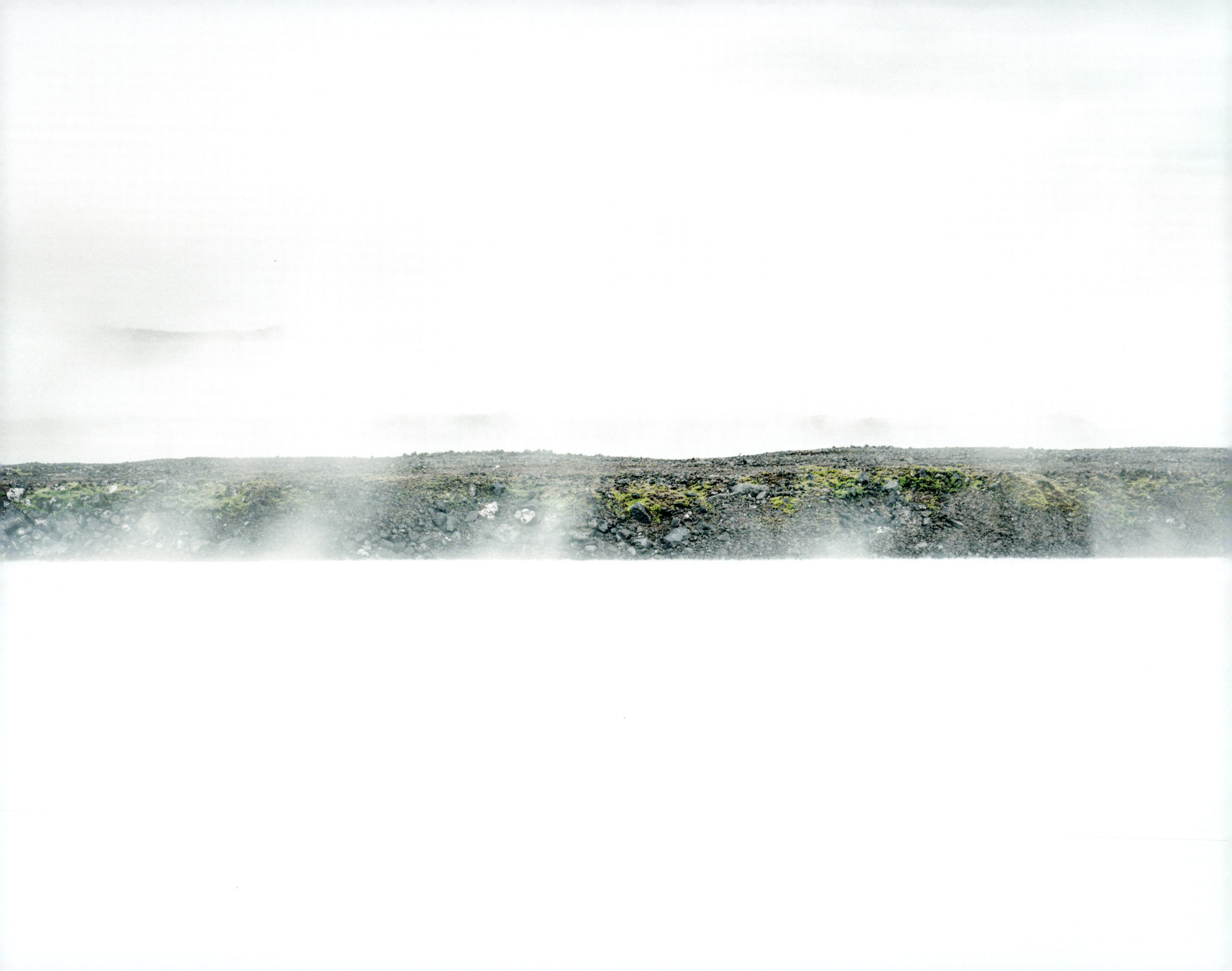

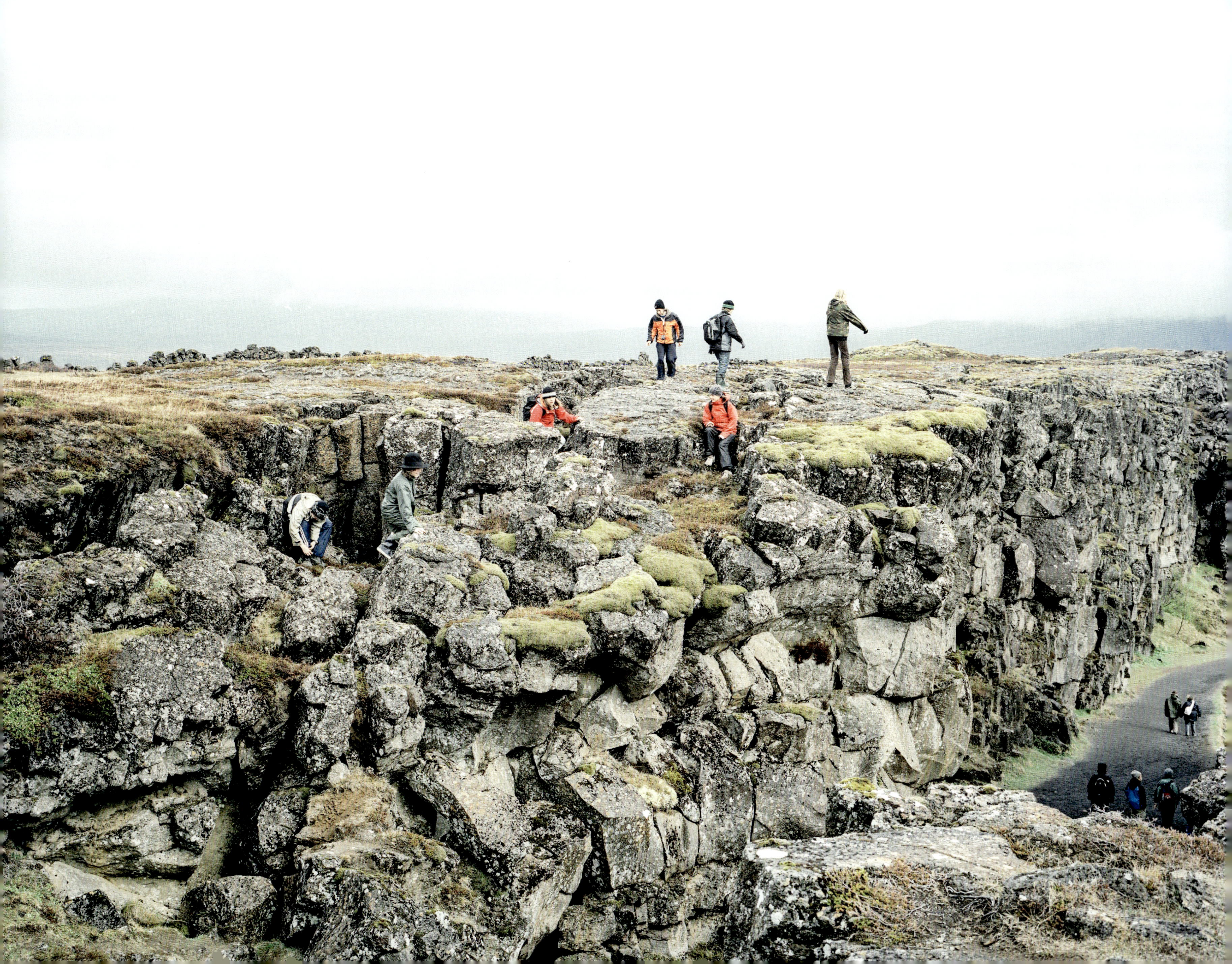

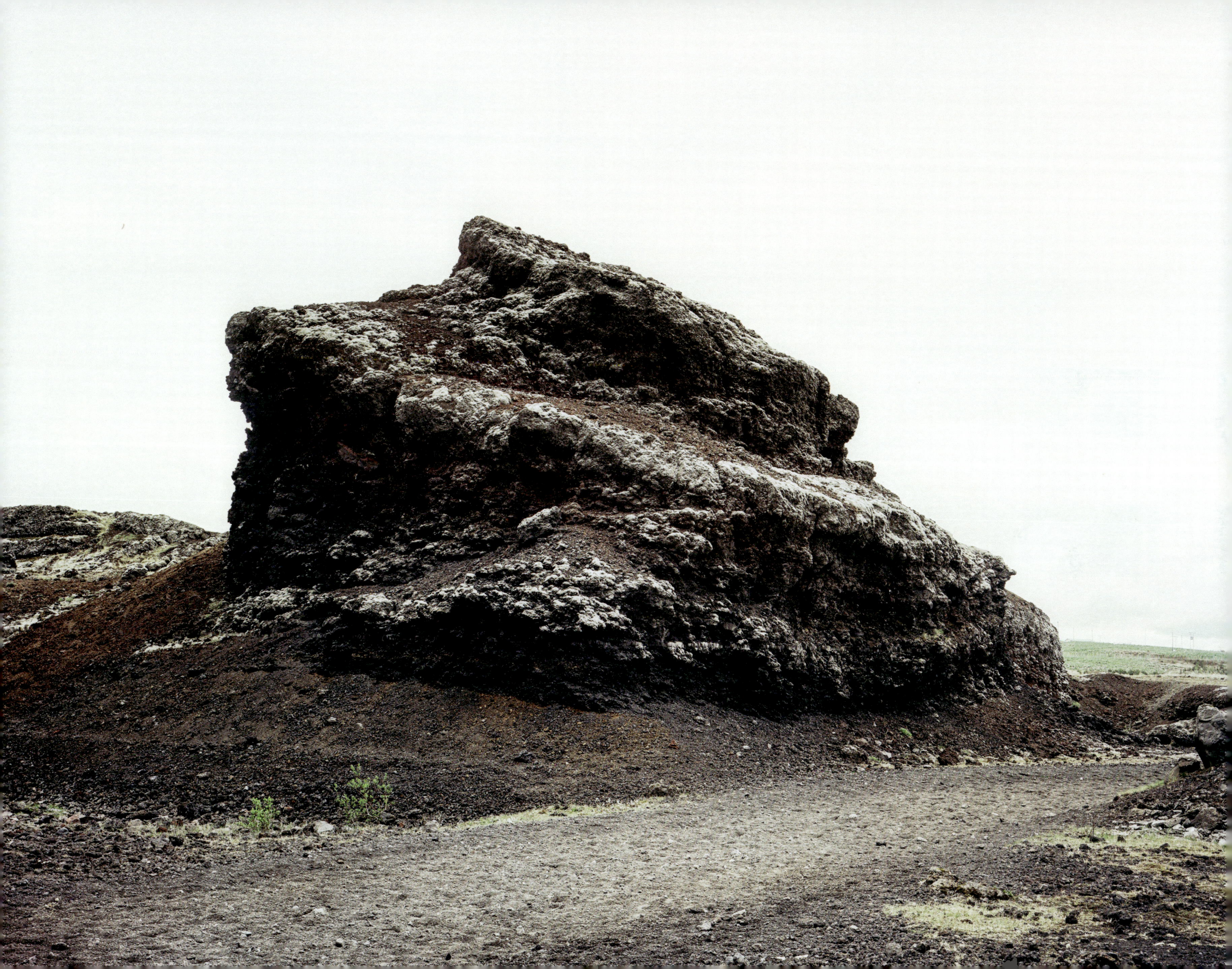

PLATE 11 | **RIFT 34**

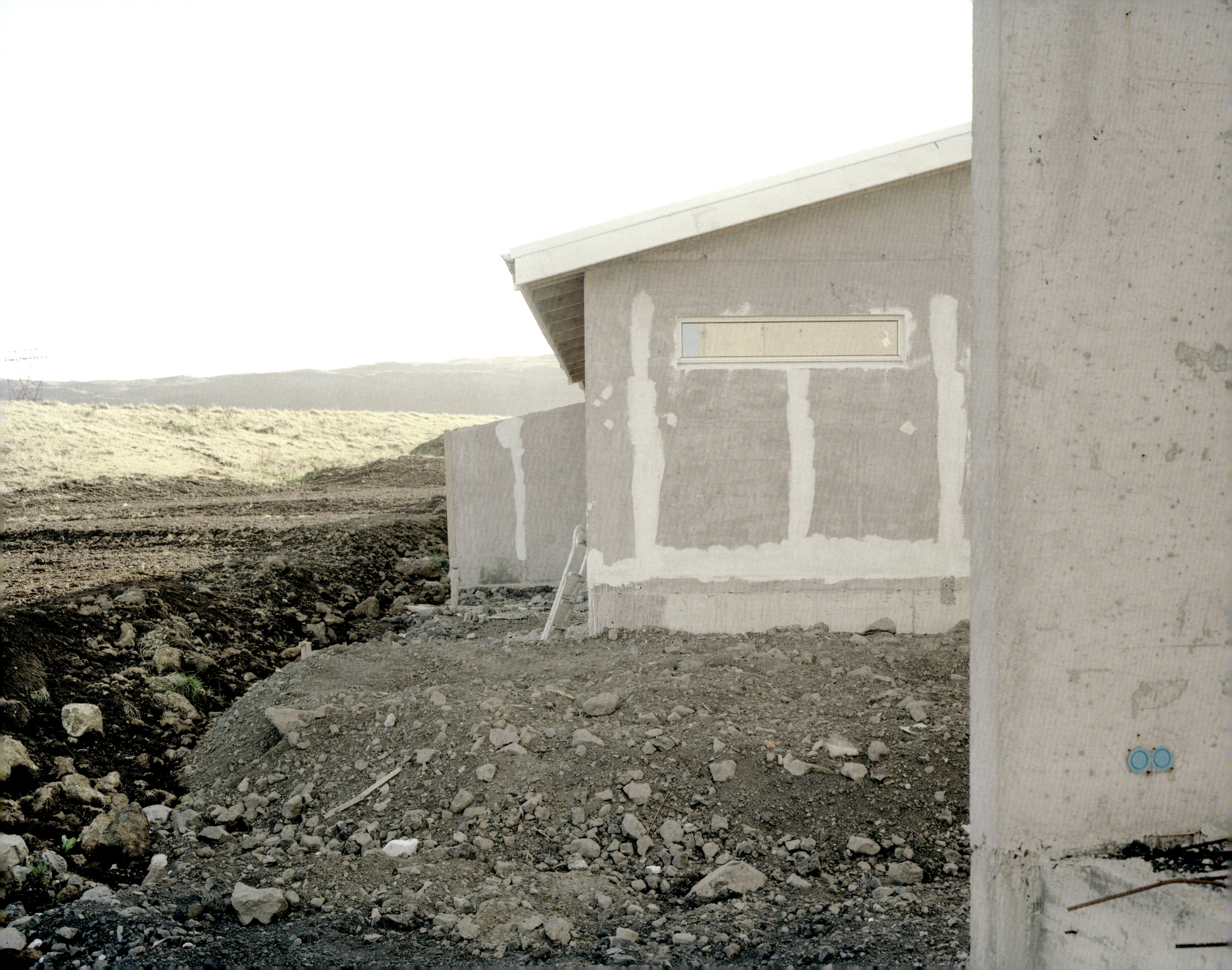

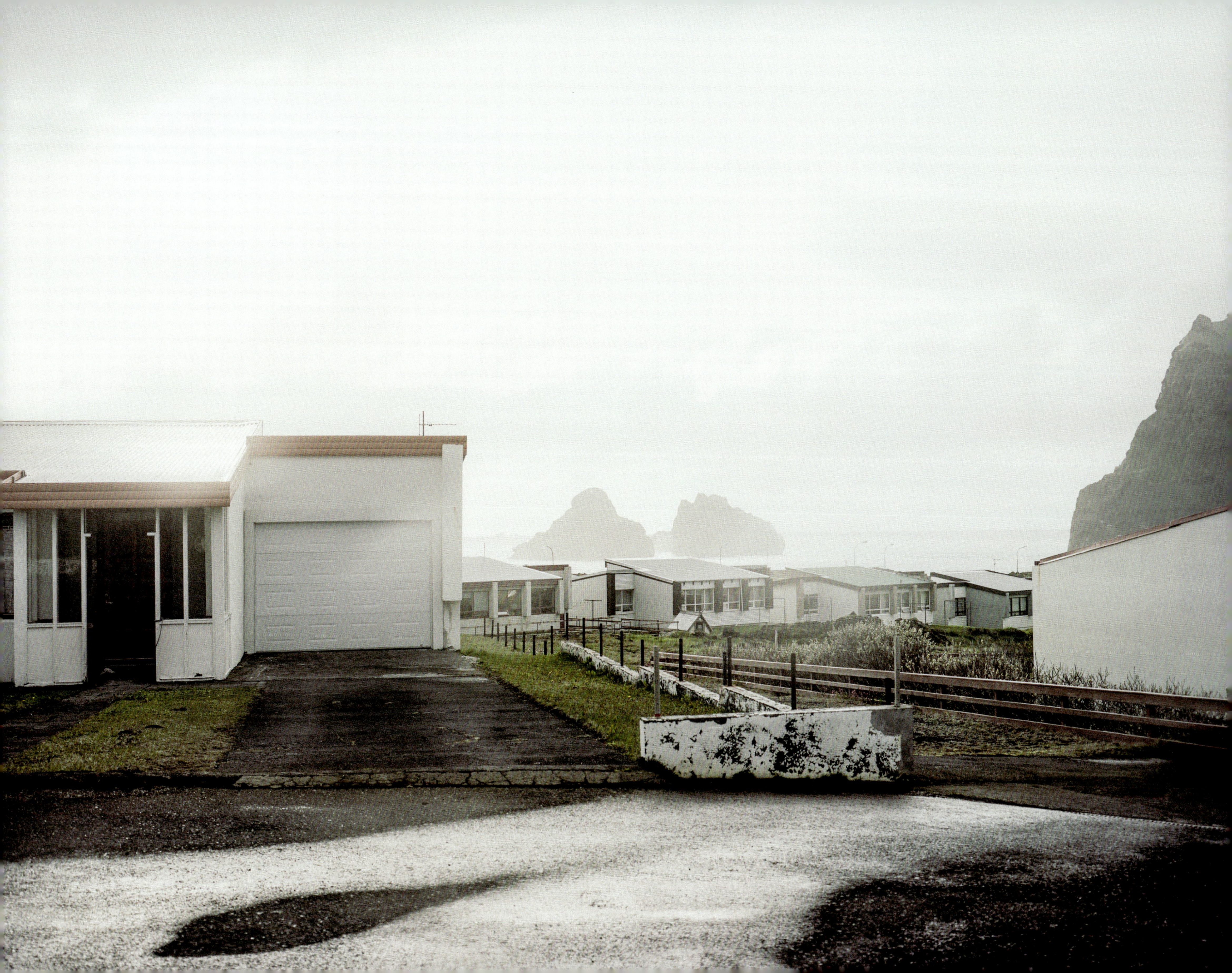

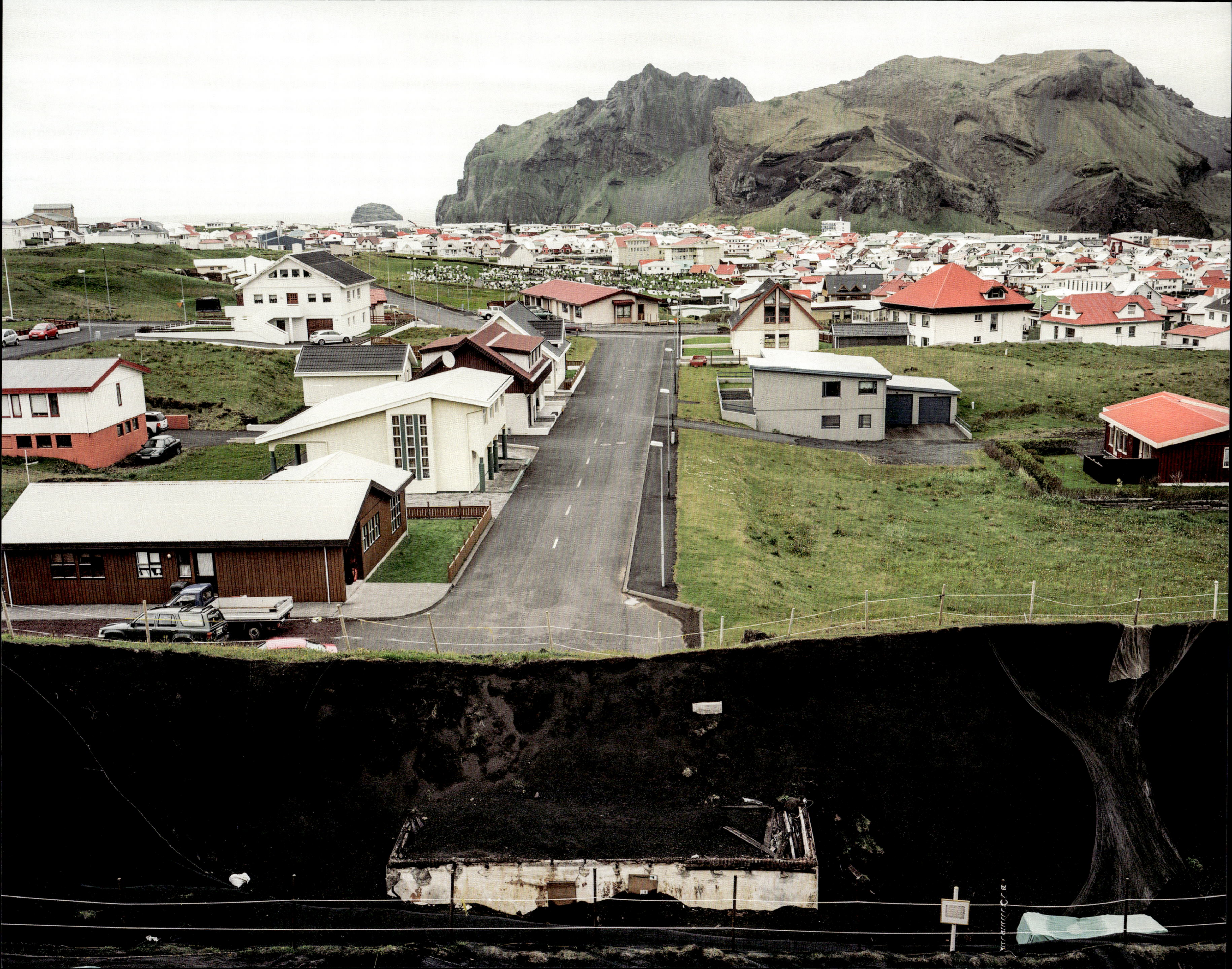

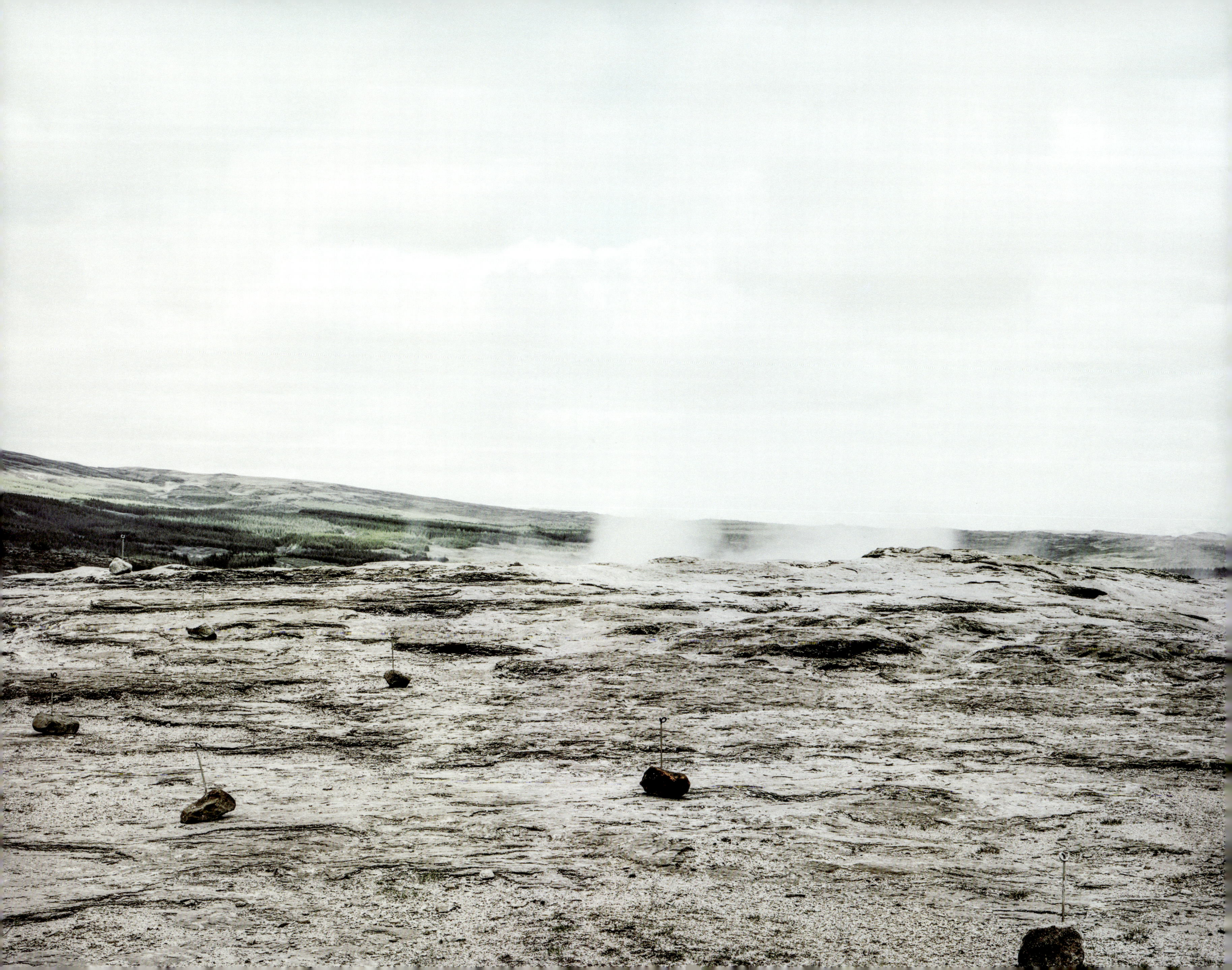

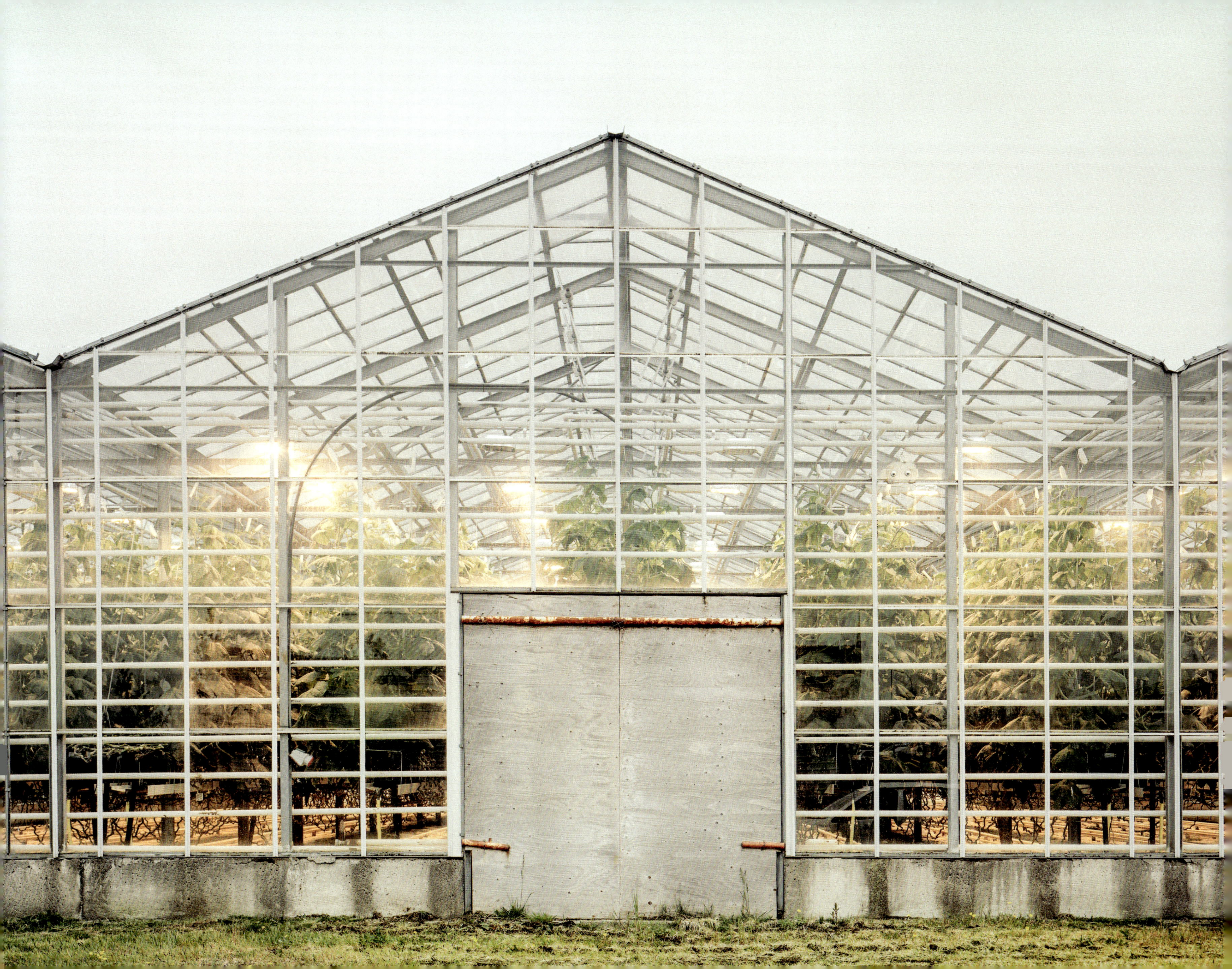

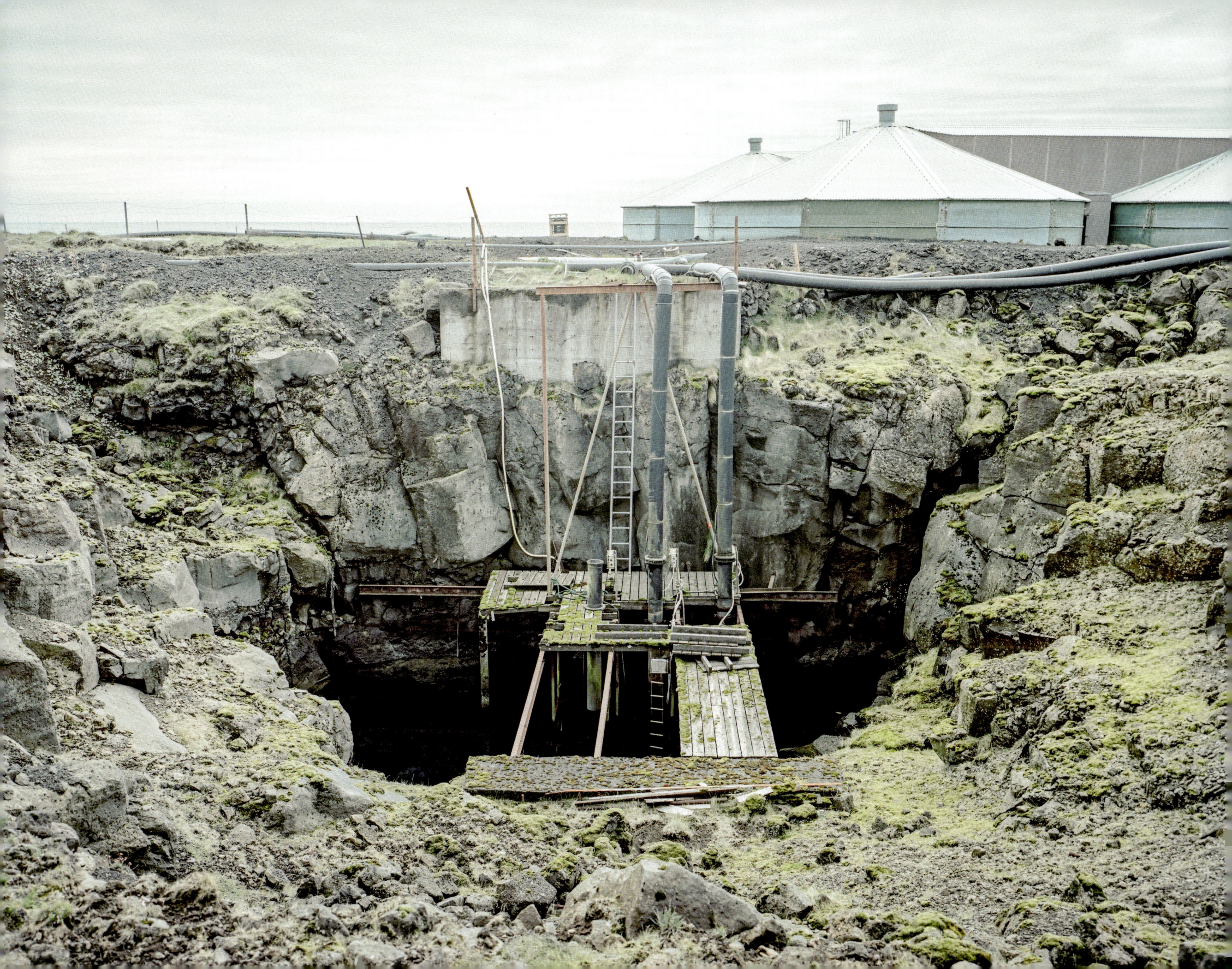

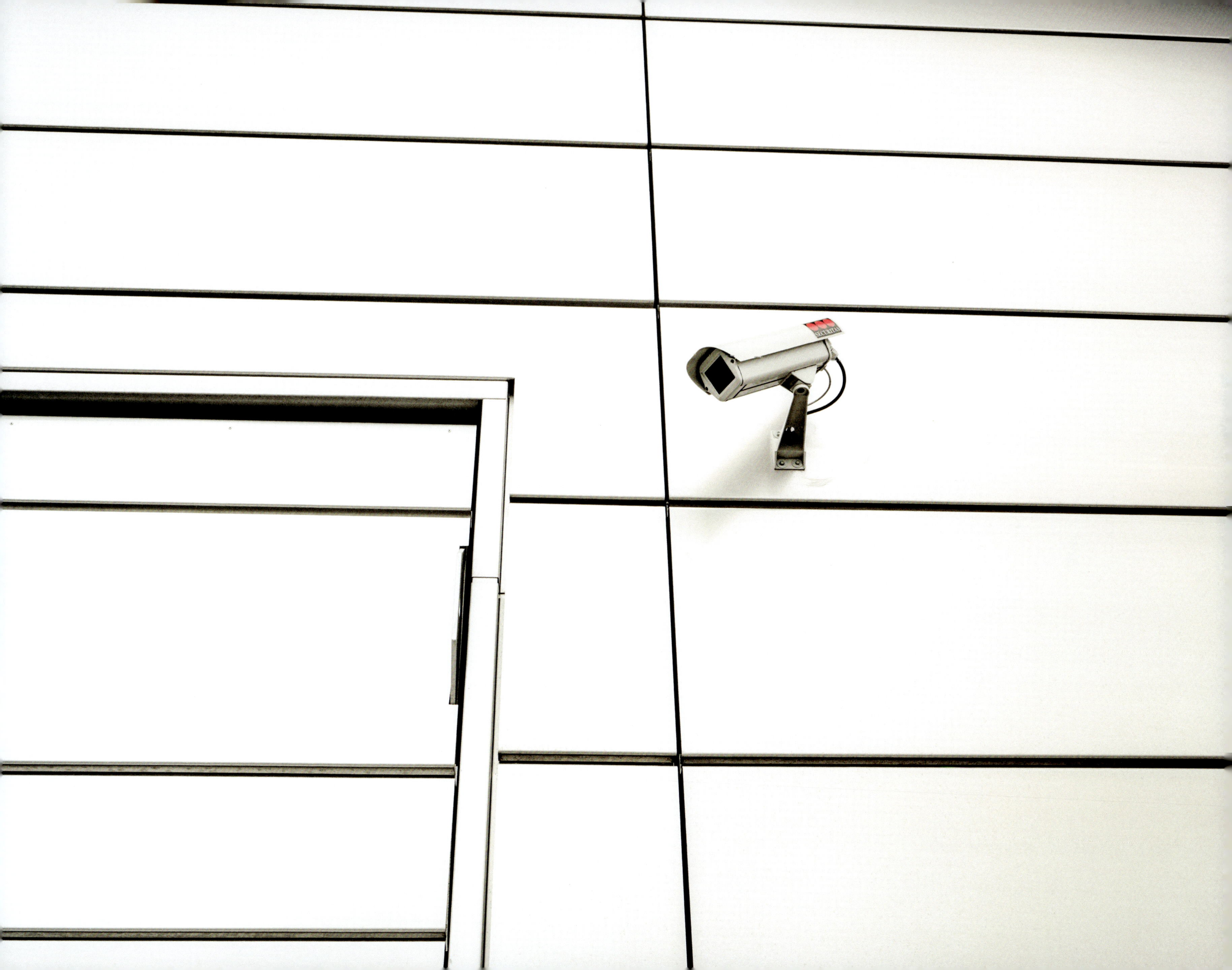

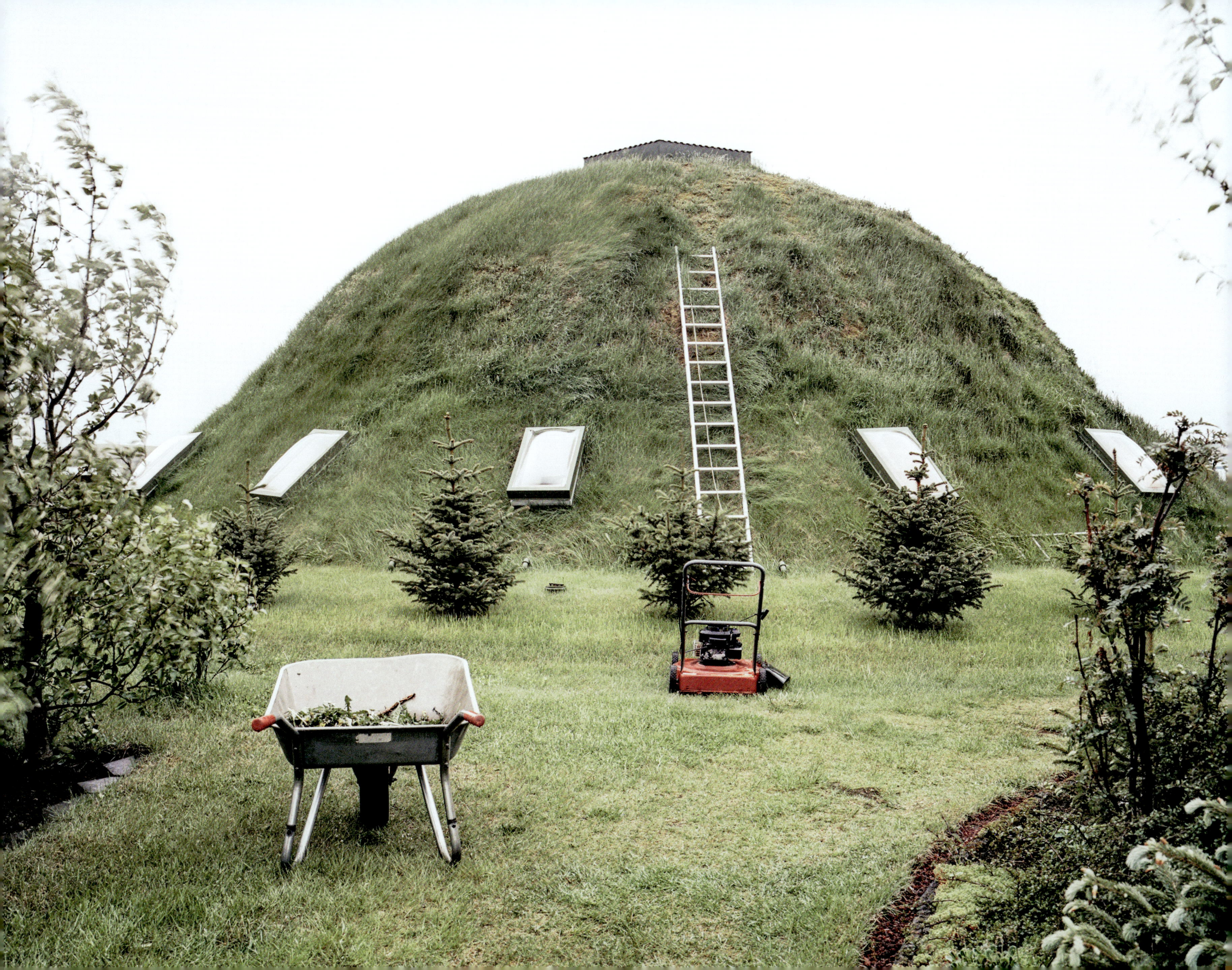

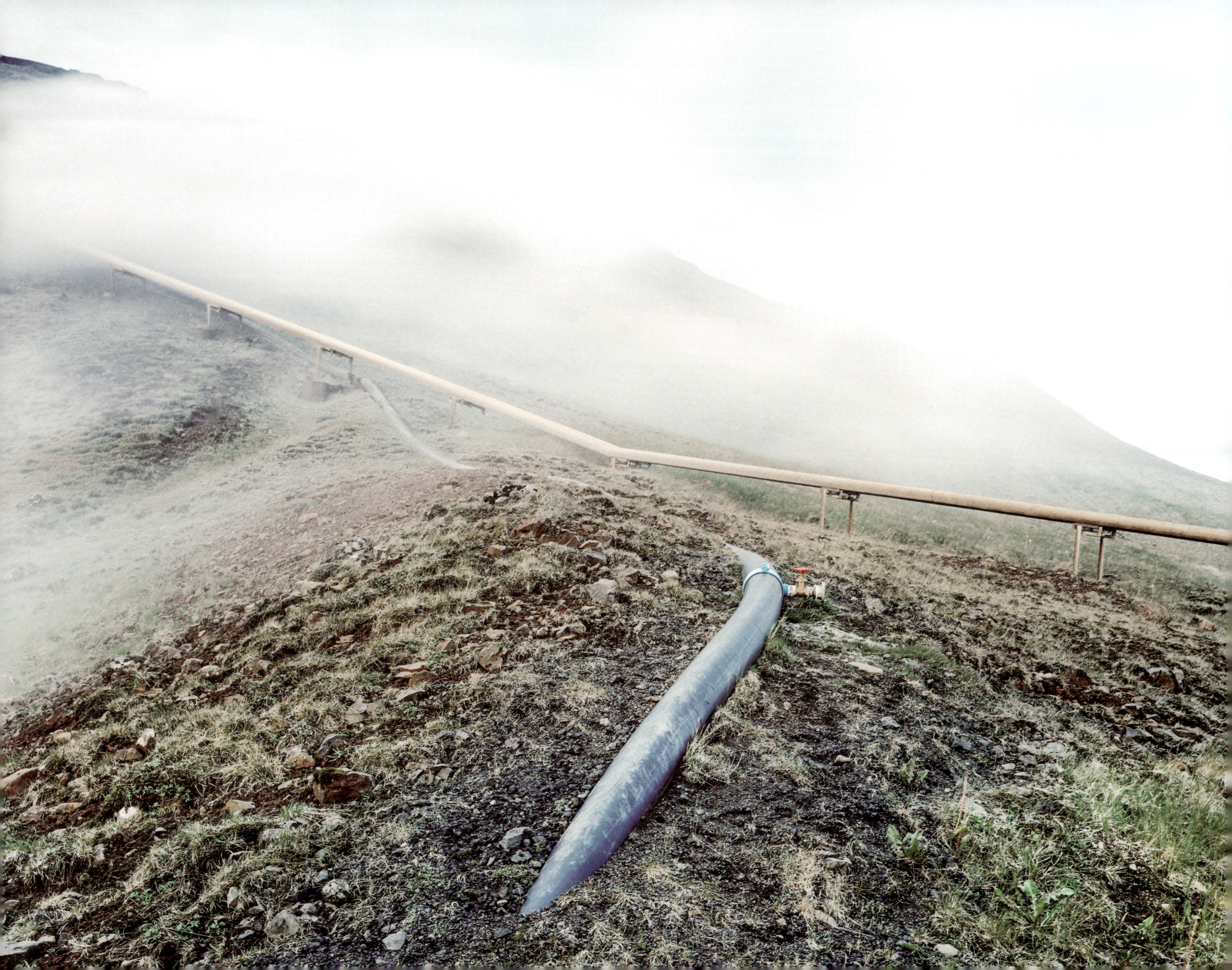

PLATE 20 | RIFT 4

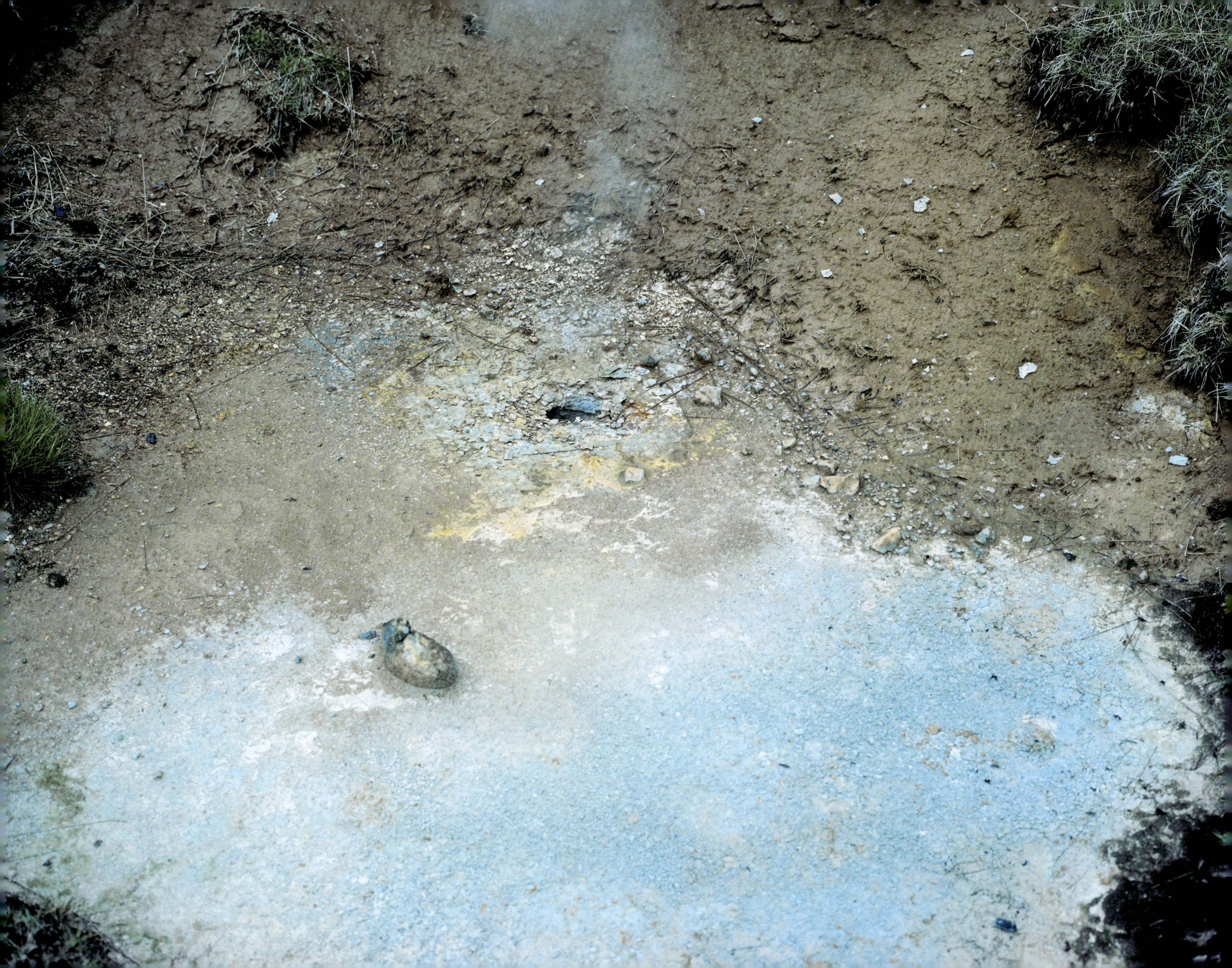

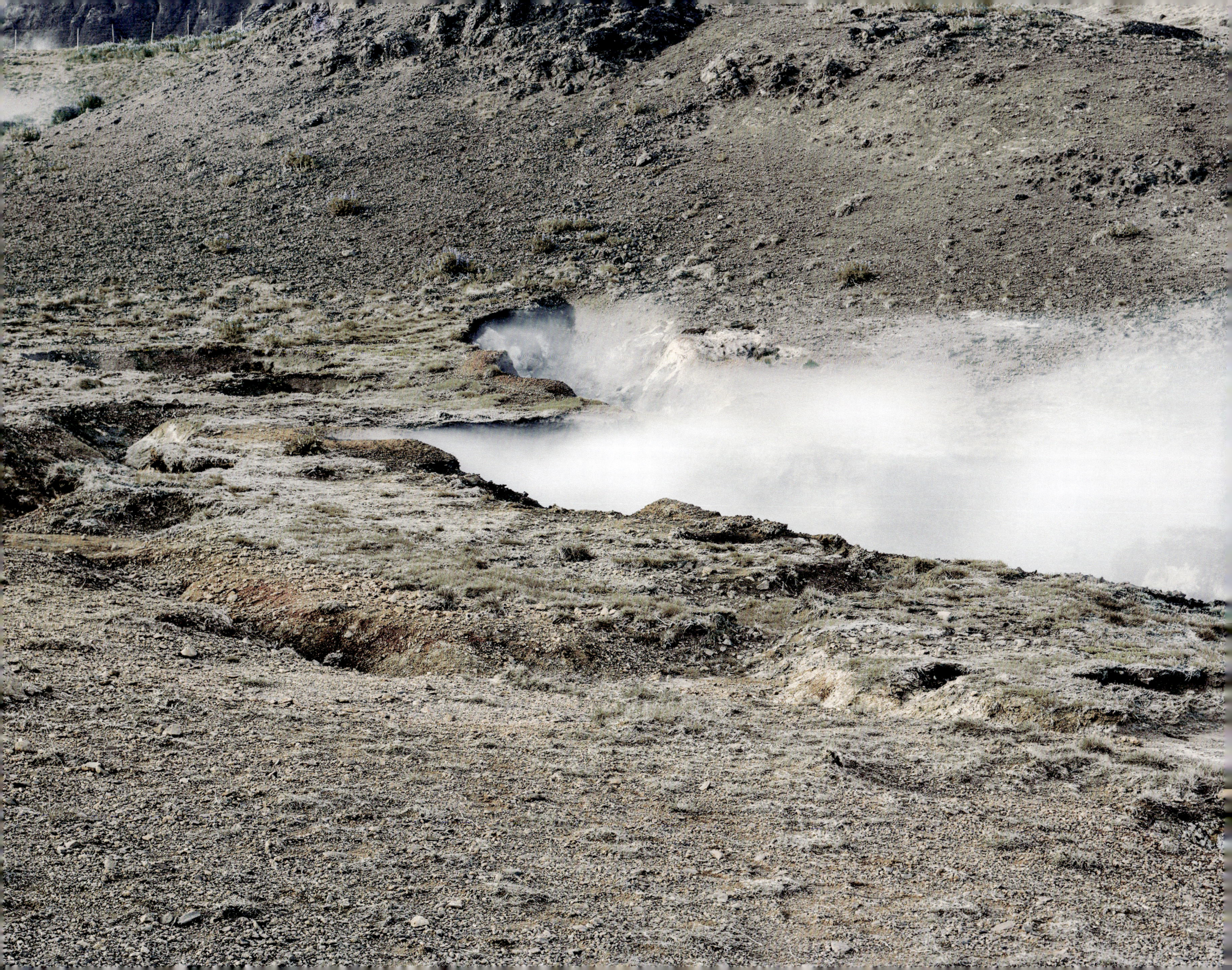

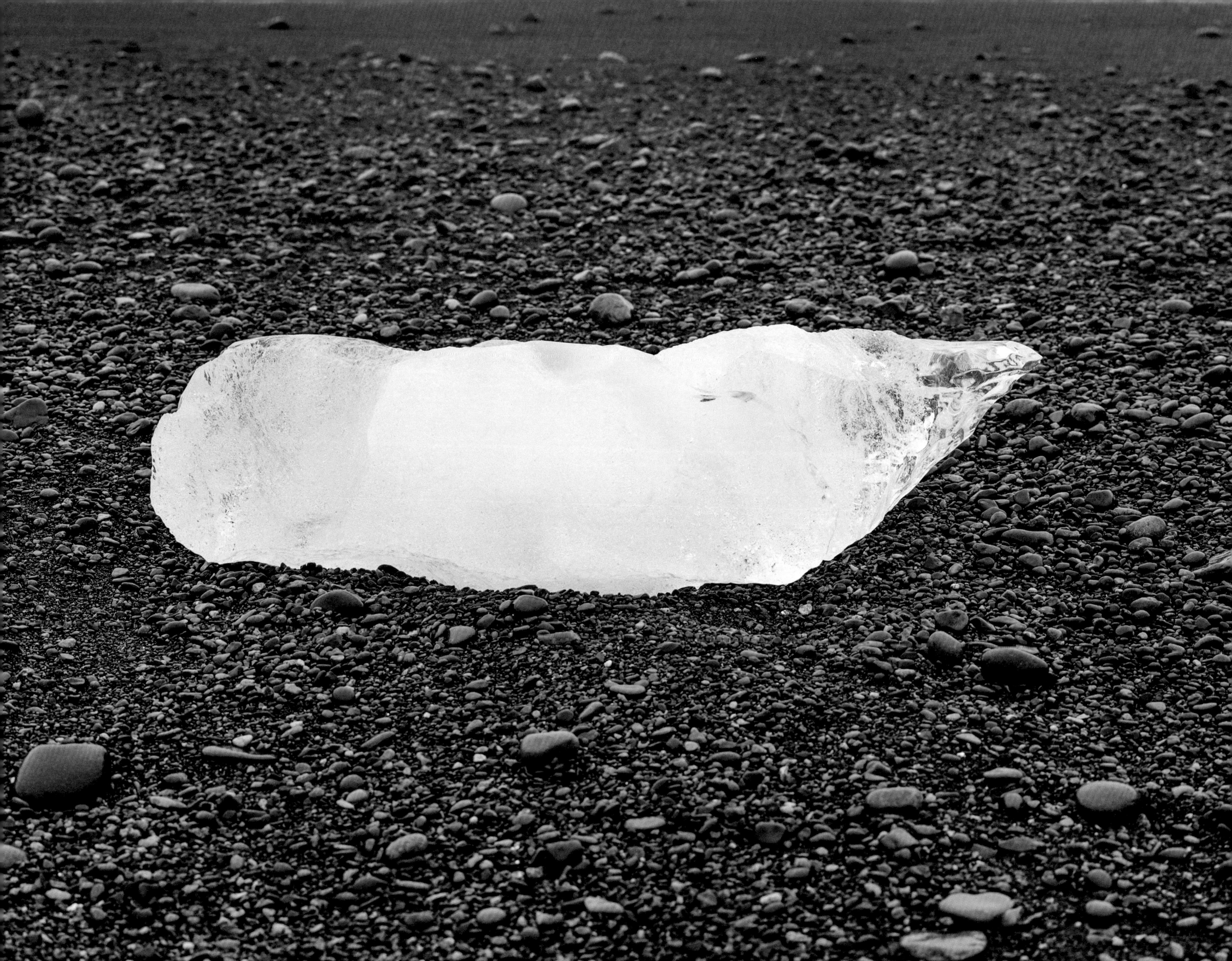

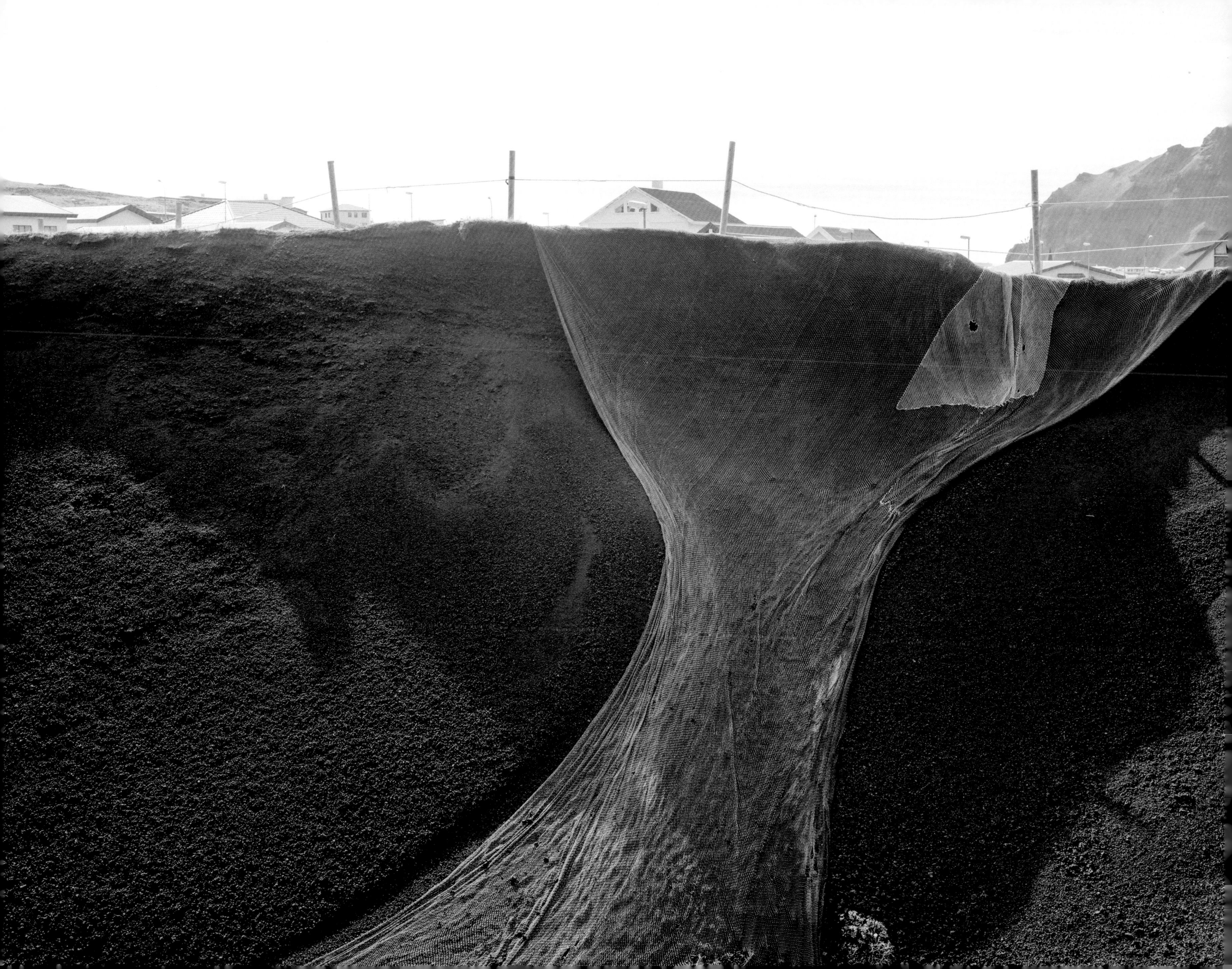

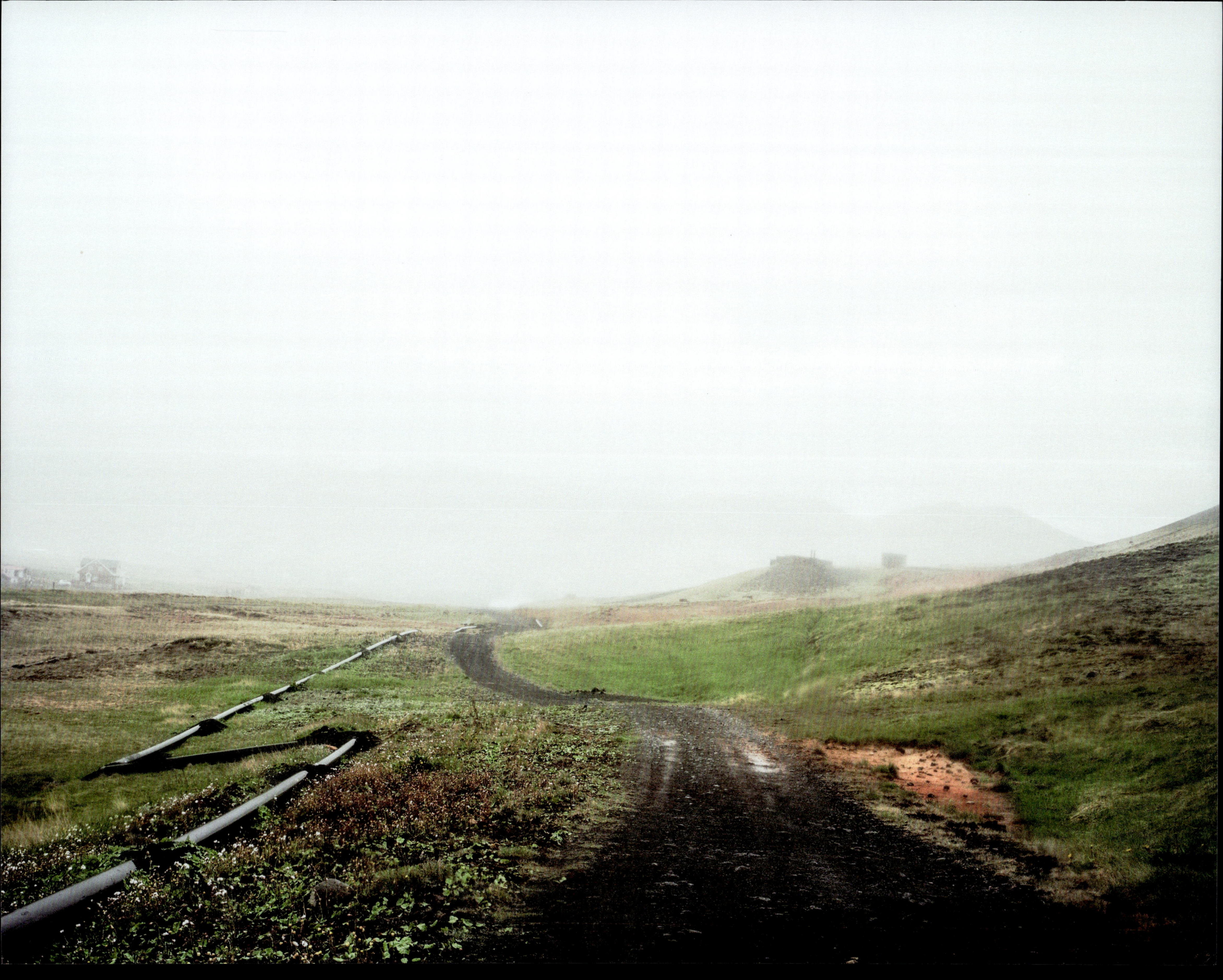

RADIUS BOOKS

227 E. Palace Ave. Suite W Santa Fe, NM 87501 t: (505) 983–4068 www.radiusbooks.org

Available through

D.A.P. / DISTRIBUTED ART PUBLISHERS

155 Sixth Ave., 2nd Floor, New York, NY 10013 t: (212) 627–1999 www.artbook.com

ISBN 978-1-942185-15-4

Library of Congress Cataloging-in-Publication Data available from the publisher upon request.

DESIGN: David Chickey

PROOFREADING: Miranda Pennington

PRE-PRESS: John Vokoun

Printed by Editoriale Bortolazzi-Stei, Verona, Italy

MARION BELANGER

RIFT | FAULT

IT IS WHAT IT IS

by Lucy R. Lippard

RIFTS AND FAULTS OFFER A POIGNANT METAPHOR for our times, as we balance on the edge of the Anthropocene and peer off into the abyss of climate change. Marion Belanger chooses to illuminate these meta-geologic processes by looking at the cultural landscapes occupying their surfaces. She comments on the visible and the invisible, acknowledgement and denial, examining, in the process, the "dangerous disconnect," where so-called ordinary lives play out in the shadows of potential cataclysm. Her photographs are paired to define these unstable boundaries: on the right, an image from Iceland, where the Mid-Atlantic Rift defines the eastern edge of the North American Continental Plate and its encounter with the Eurasian Plate, and on the left, an image from California, where the western edge meets the Pacific Plate along the San Andreas Fault. The elaborate double-book design allows viewers to create their own pairs as well. The interactive play both informs and entertains.

Belanger's provocative image pairs emphasize the message that quietly underlies the project. It begins (literally) with unexceptional social surfaces — two parking lots, one a somewhat precarious shelf at the end of the San Andreas Fault, and another blatantly flat and even gridded near an Icelandic rift. It ends with greater depths — steaming geothermal mud pools in Iceland and the Cape Mendocino coast, where a road sign warns of a "hazard zone." In between, life goes on. Old towns, new subdivisions, and homes are situated on and near rifts and faults, blithely ignoring recent histories — the 1973 volcanic eruption at Heimaey, the 1992 tsunami at Cape Mendocino, or, farther afield, the total rebuilding of Galveston, Texas, on the same sandbar after a 1900 hurricane, the restoration of towns on the Mississippi banks after inevitable floods, homes constructed on fire zone interfaces. The examples of willful amnesia are endless. Belanger documents crowds of tourists and locals clambering around on the rift wall at Thingvellir, site of the oldest parliament in the world, and on a beach near the fault. (I have clambered at both places, with not a thought to future or underlying geology.) Of course we all live our lives in some sort of denial — even those of us in Gaza, or in the range of U.S. drones — or we'd be trembling in fear twenty-four hours a day. Somewhere in between is a responsible consciousness that critical landscape photography can address and hope to affect.

Although climate change, the extractive industries, resource depletion, and so forth are on her radar, in this series Belanger deals less with imminent destruction (few people are willing to confront even that) than with more distant, more overwhelming events that are still harder to imagine. Her previous photographic subjects have included other endangered landscapes: the drastically polluted (now revived) Naugatuck River in Connecticut, where she was raised and became conscious of local environmental and economic contradictions (*River*, 2013–14); the Florida Everglades (*Everglades*, 2001–2004, with a new series in process); and abandoned public and private interiors, poignant comments on short-term impermanence (*Real Estate*, 2006). *Salt, Spiral Jetty, and Hunters* (2003) socially and environmentally contextualized a ghostly image of Smithson's iconic earthwork, including portraits of local bird hunters, bashful and boastful. *Earth in Flux* (1998-2001) is the closest earlier work to *Rift Fault*, focusing on geysers, fumeroles, and lava flows in Yellowstone National Park and Hawaii, with a cameo of *Spiral Jetty*. Shot in black and white, it offers an illuminating contrast to Belanger's color work, and perhaps a clue to her distinctive grayed or washed-out palette. In *Rift/ Fault*, a pale, even light blurs the distinctions between Iceland and California, bringing a geyser field into the same color range as an orange grove, or a geothermal plant closer to a parched basin near the Salton Sea, or the brown hills of the Coachella Valley in the San Bernardino Mountains to the gray planes of an unfinished stuccoed structure in Icelandic lava fields (halted by financial crisis, though the sun striking a nearby field suggests relief).

It is here that the photographer's "imaginative space" fuses with the world we all see. Her photographs are often subtle and elusive. Belanger is fond of mists, fog, and steam. She prefers mystery to melodrama. Landscape photographers learn to frame vastness, to make sense of space. Like Joe Deal and Edward Ranney, among others, Belanger can make what at first glance looks like nothing look like something we should have noticed. Once we get it, understatement itself becomes meaningful. For instance, the striking image of an excavation site at Heimaey, where 400 homes were buried in the 1973 eruption of Eldfell, silently reaches across the continental plate to a red board fence bordering a Daly City neighborhood of modest houses lying directly on the San Andreas Fault, which now and then causes lethal landslides. The Icelandic excavation is a yawning black maw, an unavoidable gateway into the bowels of the earth, recalling the epidemic of unpredictable sinkholes in the southern United States. In another pair, a vast San Bernardino subdivision disappears in the smog; an Icelandic hydrologic power station, almost invisible above a cloud of steam, also feeds an immensely popular tourist destination—industrial-scale geothermal baths at the nearby Blue Lagoon.

In her artist's statement, Belanger notes that she looks for "visual tension" and "moments of quiet anticipation" to heighten the "uneasy relationship between geologic force and the limits of human intervention." Interfaces and borders make the point. Images of failed or failing California real estate provide an almost ludicrous parallel to the shifting plates. The cultural landscape rarely exposes the geological secrets beneath it until viewers begin to put the pieces together (literally, given the book's design, which offers creative impermanence).

A secondary theme might be a quiet commentary on the industrialization of nature. The gigantic structures of a southeastern Icelandic fish farm bear no resemblance to adjacent seas and old fishing shacks. Compared with Mussel Rock Park, a popular California birding site that claims to be "the closest point to the epicenter of the 1906 San Francisco earthquake." Natural and unnatural fecundity are again evoked by a flourishing Icelandic greenhouse, geothermally heated throughout the winter, juxtaposed with a cactus farm framed by wind turbines. The contradictions lie in the content as well as in the visuals. We discover that a bleak but green Icelandic landscape with road and pipelines is actually a thriving farming area thanks to geothermal steam, while in California we find a tier of dilapidated signs for unsalable lots for sale near the Salton Sea, where agriculture is stymied by the increasing salinity of this inland lake with no outlets.

The frequency of human-made structures (and far fewer humans) in a photographic series devoted to the grandest natural forces in the earth's history extends quotidian views of the current built environment into a vaster time scale. Landscape is almost by definition context — time and space. We know nothing is permanent, but the concept of temporary is relative. When I studied geology in college more than fifty years ago, the professor had been a philosopher, and he brought geological time in the Connecticut River valley alive as few scientists could. Belanger's double images compare times — past and future — as much as near and far spaces. The answer to Kevin Lynch's classic question, "What Time is this Place?" takes a giant leap from a birthing earth to contemporary trivia.

Ironies abound. A geothermal borehole housed in a geodesic dome providing power in Iceland reflects on its grim counterpart — a seismometer structure in Parkfield, California, "Earthquake Capital of the World," where the fault plate moves constantly. Consider the positive power of Iceland's rifts (85% of the small nation's energy comes from geothermal) accompanied by the negative threat of volcanic eruptions. Then consider the negative power of the San Andreas Fault, where one day all hell is going to break loose, and the positively idyllic coastal settings that lure people to nest in its hazard zones.

Abandonment (or waste) is another implied subtext, evoking Alan Weisman's haunting book *The World Without Us*. A lava-buried house with an exposed chimney is contrasted with brandnew housing abandoned near San Bernardino, where a lone dead palm tree stands silhouetted against a bleak stony vista. Similarly isolated, stick-like young trees are protected as they struggle to grow in an Icelandic pseudocrater field (formed by lava flows from an eruption in the tenth century), juxtaposed against a stressed-out, windblown patch of palms and cactuses near one of the fastest growing cities in California. The lack of forests in Iceland and desert California informs the pairing of a turf-covered geodesic dome and home on a barren residential lot. (Turf and driftwood stand in for scarce timber in Iceland; the island's birch forests were decimated by early settlers, and newer trees are stunted. A local joke: What do you do if you get lost in an Icelandic forest? Stand up.)

Sometimes Belanger's juxtapositions are formally witty, such as an odd lava hill next to the blank back side of a polygonal sign at Desert Hot Springs. Another striking duo pairs a near abstract image of icy blue and white — a volcanically heated public pool — next to a neatly laid concrete block wall — a bulwark against inevitable earthquakes. (When Belanger went back to rephotograph the pool, it had vanished.) And in another, the flat black horizon of a lava field against a whitish sky is paired with a seascape of bleached water and sand and misty black rocks at Cape Mendocino Triple Junction — "one of the few places in the world where three gigantic tectonic plates meet, creating a seismically active zone."

We are familiar with volcanism, tsunamis, and earthquakes (some recently triggered by fracking), but the unfamiliar vocabulary of tectonic plate theory is evocative, almost performative, suggesting sculptural strategies like those that inspired Robert Smithson in the 1960s: Slab Pull and Ridge Push, Subduction Zone, Mantle Convection, Active and Passive Rifting, Blind Faults, Continental Lithosphere, Gravity-Driven Extensional Faults. Belanger has educated herself in the science, but the viewers of these photographs need only be aware of the grandeur and potential violence, the minute but unceasing movements undermining the edges of our continent. Her research describes, while light on the land defines.

In a growing field of critical landscape photographers, the conceptual clarity and understated execution of *Rift/Fault* stands out. I look for content: engagement with location and issues, photographers who insist on context, social narrative, and meaning. These goals are usually facilitated by a narrative seriality or aesthetic sequencing that seems suited to representing mobility and change. I like to think of a photograph as a *field* rather than an *artifact*, suggesting layers and periphery. Though we no longer see photography as "truth," since digital technology has made lies so easy, it still conveys firsthand experience like no other medium, pointing out what we haven't seen — so long as captions tell us what we are seeing.

Iceland, with its geothermal energy, hydroelectric, and wind farms (less than 0.1% of its power comes from fossil fuels) is a model for the rest of the world, especially as California burns. But it's a trade-off. Unlike climate change and so many other vital ecological issues, there is nothing we can do about the great beast stirring restlessly beneath our feet, beneath our lives. It is what it is. But at the same time our very powerlessness can remind us of what we can change. *Rift/Fault* is not only a work of art, it's a wake-up call.

Fault images were made along the San Andreas Fault in California.
Rift images were made along the Mid-Atlantic Ridge in Iceland.

1. **Fault 52**, 2012
    Point Delgada is the end point of the San Andreas Fault proper. Thereafter the fault is
    splintered, and moves from sea to land until it finally departs land at Cape Mendocino.

    Rift 31, 2007
    A parking lot near Hveragerði on Route 1.

2. **Fault 56**, 2012
    Mussel Rock Park in Daly City is located at the base of a cliff overlooking the Pacific
    Ocean, just south of San Francisco. The San Andreas Fault departs land here before
    returning to shore near Stinson Beach.

    Rift 8, 2006
    The Leirhnjúkur area is a central volcano with steaming hot lava fields in the area of Mývatn
    and close to Reykjahlíð in the northern central region of Iceland. The Apollo 11 crew trained
    in the barren, steaming landscape here for their impending moonwalks.

3. **Fault 22**, 2012
    In San Bernardino, during the economic crisis, many real estate projects like this inactive
    housing development were halted, stalled, or abandoned.

    Rift 25, 2007
    Heimaey is a fishing island located in the Vestmannaeyjar archipelago. It is the largest and
    most populated island off the Icelandic coast. On January 23, 1973, the volcano Eldfell erupted
    without warning and volcanic ash destroyed over 400 homes. .

4. **Fault 24**, 2012
    Wind turbines, Desert Hot Springs. The San Gorgonio Pass is one of the windiest places in
    southern California.

    Rift 22, 2006
    Originally, birch trees covered 40% of Iceland, but after the Vikings settled during the 9th
    century the island was completely deforested. Trees are now thriving, mainly due to climate
    change and warmer temperatures. These young trees are growing in the Landbrotshólar
    pseudocrator field.

5. **Fault 18**, 2009
   The Salton Sea is a shallow, saline, rift lake located directly on the San Andreas Fault in California's Imperial and Coachella valleys. It is one of the world's largest inland seas and lowest spots on earth at -227 below sea level. Landlocked, with consistent rising salinity, the Salton Sea is now considered an ecological disaster.

   Rift 23, 2006
   Geothermal borehole structure at the Nesjavellir Power Station. Nesjavellir is the second largest geothermal plant in Iceland. It produces municipal hot water and electricity to roughly 85% of all houses in Iceland.

6. **Fault 15**, 2009
   San Bernardino is in a high-risk earthquake zone. Recent scientific evidence suggests that a large earthquake along the southern San Andreas Fault is overdue. Most at risk is the 100-mile segment that cuts through Palm Springs, San Bernardino, and Riverside.

   Rift 18, 2006
   Located in the inlands of Árnessýsla, Laugarvatn shares its name with the shallow, heated from underground lake where the town is sited. Almost every town in Iceland has a volcanically heated public pool, which serves as an unofficial town social center, and plays an important role during the long, dark winters. This pool, formerly heated naturally, has since been replaced with a modernized hot springs spa.

7. **Fault 16**, 2009
   San Bernardino, located 60 miles east of Los Angeles, is the 17th largest city in California.

   Rift 62, 2011
   At the Svartsengi Power Station in Reykjanes Peninsula, warm, surplus mineral-rich water from the power plant feeds the Blue Lagoon next door. The geothermal plant pipes hot water to more than 21,000 households, and is the only heating system in the local district.

8. **Fault 55**, 2012
   This area, called the Mendocino Triple Junction, is one of the few places in the world where three of the gigantic plates meet. Along most of California's length, the San Andreas Fault defines the boundary between the Pacific Plate to the west and the North American Plate to the east. The Pacific Plate slides horizontally in a north northwesterly direction, and then the fault splinters and fragments as it moves to sea and joins the Mendocino Triple Junction. This massive fault has been responsible for eruptions at Crater Lake and Mount St. Helens.

   Rift 12, 2006
   Black ash and lava field near the Hekla Volcano in south Iceland. Hekla, one of Iceland's most active volcanoes, last erupted in 2000. The volcano's frequent, large eruptions have covered much of Iceland with volcanic tephra.

9. **Fault 50**, 2012
    Camping Beach, Route 1 north of Point Reyes. Traveling north, the San Andres Fault travels
    out to sea at Daly City, and from then it moves back and forth from land to water.

    Rift 1, 2006
    Thingvellir is a UNESCO World Heritage Site because from 930 AD to 1798 it was the
    location of Iceland parliament, the oldest in the world. It is also a National Park
    because it is situated on the tectonic plate boundaries, which created the canyon
    Almannagjá between them.

10. **Fault 37**, 2012
    Desert Hot Springs is famous for its hot mineral spas, which originate from an underground
    reservoir. The fault runs directly through the town.

    Rift 11, 2006
    The Rauðhólar ("red hills"), outside of Reykjavik, are remnants of a cluster of pseudocraters
    that are about 5200 years old. Pseudocraters form when lava flows into rivers or wetlands.
    When the very hot lava comes in contact with water, heavy explosions occur from the boiling
    water and hot steam. American soldiers used soil from the pseudocraters to build an airstrip
    during WW II. Today, the area is part of a nature park.

11. **Fault 30**, 2012
    The Coachella Valley is a desert valley near the San Bernardino Mountains that extends for
    approximately 45 miles in Southern California, to the northern shore of the Salton Sea. The
    San Andreas Fault runs through this Valley.

    Rift 34, 2007
    New construction in the lava fields of Hveragerdi, at the tail end of the Nordic housing
    bubble, and just before the financial crash in late 2008. The Icelandic financial crisis
    involved the default of all three of the country's major privately-owned commercial banks.
    This collapse was the largest experienced by any country in economic history.

12. **Fault 23**, 2012
    Night view of a neighborhood built upon the San Andreas Fault zone in Southern California.

    Rift 42, 2011
    A neighborhood in the town of Heimaey in the Vestmannaeyjar archipelago.

13. **Fault 10**, 2008

Daly City, located along a cliff overlooking the Pacific, is just south of San Francisco. The city is the largest within San Mateo County. The San Andreas runs along the base of the cliff, an active landslide that takes down homes, dirt, and rock.

Rift 26, 2007
A volcanic excavation site on Heimaey.

14. **Fault 26**, 2012
The Salton Sea covers up the southern end of the San Andreas Fault. There are no water outlets for the inland lake. This, and agricultural runoff, are responsible for the ever-increasing salinity of the water. Property values have plummeted, and tilapia, the only surviving fish, wash ashore and bake under the hot sun. A combination of soil from the dried up lake and a toxic mix of chemicals, salt, and metals creates dangerous dust storms when the wind blows.

Rift 63, 2011
Markers at the original (Great) Geysir, on the slope of Laugarfjall in Haukadalur, in the southwestern region of Iceland. Geysir has been active for approximately 10,000 years, with eruptions of boiling water reaching up to 70 meters in the air.

15. **Fault 53**, 2012
In 1992, a tsunami was generated by a magnitude 7.1 earthquake near Cape Mendocino. This is where the San Andreas Fault finally leaves the land. Here the Pacific plate gives way to the Juan de Fuca plate, a remnant of the ancient Farallon plate that is moving against the North American continent and being forced beneath it—an area known to seismologists as the Cascadia Subduction Zone (CSZ).

Rift 48, 2011
Like Hveragerdi, Reykholt is a greenhouse village that uses geothermal steam for heat, even during the winter months.

16. **Fault 28**, 2012
This cactus farm, near Desert Hot Springs, is indicative of how California municipalities are encouraging residents to replace green lawns and flowering annuals with indigenous desert plantings. This desert region has an annual rainfall amount of less than 5 inches and ongoing drought. Seen in the background is the San Gorgonio Pass Wind Farm, the oldest of three wind farms in California. There are 4,000 wind turbines in a 70-square-mile area that provide enough electricity to power Palm Springs and the entire Coachella Valley.

Rift 32, 2007
The fishing industry is one of the main pillars of the Icelandic economy, and includes fish farms like this one. Marine products have historically been the country's leading export items and the seafood industry is based upon sustainable practices.

17. **Fault 45**, 2012

Called the Earthquake Capital of the World, Parkfield, with a population of 18, is the most seismically monitored location in the world for earthquake activity. The San Andreas Fault is constantly moving here, and every 20-30 years strong earthquakes occur. The last large-magnitude earthquake in 2004 registered 6.4.

Rift 27, 2007

Geothermal power plants, like this one located on the Reykjanes peninsula, produce nearly 20% of the country's electricity and geothermal heating supplies nearly 90% of the country's domestic heating and hot water requirements. Nearly all the rest comes from hydroelectric generation, with less than 0.1% from fossil fuels.

18. **Fault 11**, 2008

In Daly City, homes were built in the 1950–60's when the scientific community embraced an acceptance of plate tectonics as a geologic fact. The San Andreas runs along the base of the cliff here.

Rift 21, 2006

In Iceland, there is a long tradition of turf-covered dwellings, like this geodesic dome near Hella. This was due to the severity of the weather, and the superior insulation qualities of turf, as well as the scarcity of wood on the island.

19. **Fault 6**, 2008

Mussel Rock Park is built upon a landfill, at the closest point to the epicenter of the 1906 San Francisco earthquake. The houses built upon the edge of the cliff are at high risk for sliding down due to mudslides and earthquakes.

Rift 20, 2006

Located near the historic site of Thingvellir and the Hengill Volcano, Nesjavellir is the second largest geothermal plant in Iceland. It produces municipal hot water and electricity to the capital city of Reykjavik.

20. **Fault 43**, 2012

As evidenced by this cracked retaining wall in Hollister, the Calaveras Fault is slowly splitting the city in half, at a rate of half an inch per year. Hollister is well known among geologists because it portrays one of the best examples of aseismic creep in the world.

Rift 4, 2006

South of the Krafla caldera region is the steaming geothermal area of Hverir, with boiling mud pools.

21. **Fault 19**, 2012
    While housing developments have replaced many of the orange groves in San Bernardino
    County, there are still some among the suburban landscape.

    Rift 64, 2011
    Hveragerði is a town and municipality in the south of Iceland located about 30 minutes east
    of Reykjavík. The town sits upon a lava field, and there are thermal springs throughout it. On
    May 29th, 2008, an earthquake shook the region and a new fissure opened up on a hillside
    above the town. This ground has become unstable and dangerously hot.

22. **Fault 36**, 2012
    The abundance of litter, such as these plastic bags in Desert Hot Springs, is not seen in Ice-
    land. Shoppers there bring their own bags, or if not, stores charge to provide them.

    Rift 12, 2006
    Jökulsárlón is a large glacial lake in southeast Iceland, situated on the head of the
    Breiðamerkurjökull glacier. As the glacier retreated from the Atlantic Ocean the lake began to
    form. Global warming has accelerated the melting of the ice, and the lagoon has grown larger.
    This iceberg remnant travelled from the lagoon to the black sand edge of the ocean.

23. **Fault 3**, 2008
    Wallace Creek, Carrizo Plain National Monument. The Carrizo Plain is in southeastern San
    Luis Obispo County, 100 miles northwest of Los Angeles. It is the largest native grassland
    remaining in California. The San Andreas Fault cuts across the plain. Wallace Creek is famous
    for the 7.9 magnitude Fort Tejone Earthquake of January 9, 1857, when the town was sheered
    in half, offsetting the creek bed 30 feet.

    Rift 29, 2007
    Volcanic excavation, Heimaey. On January 23, 1973, the volcano Eldfell erupted without warn-
    ing on the fishing island of Heimaey. The eruption was slow moving and lasted until July 3rd.
    During the eruption, half of the town was destroyed and the island expanded by a third.

24. **Fault 54**, 2012
    Rain and fog, Shelter Cove Road. This road descends down from the King Range into Shelter
    Cove and Point Delgada, where the San Andreas Fault proper ends.

    Rift 51, 2011
    The Varma River and geothermal pipes, Hveragerði. This river runs through the steamy,
    geothermally active valley known for its greenhouses and hot springs.

For Valarie Rose and John Forrest

This booklet is a companion to *Rift | Fault*
published by Radius Books in 2016.

RADIUS BOOKS
227 E. Palace Ave. Suite W Santa Fe, NM 87501
t: (505) 983–4068
www.radiusbooks.org

ISBN 978-1-942185-15-4

Printed by Editoriale Bortolazzi-Stei, Verona, Italy